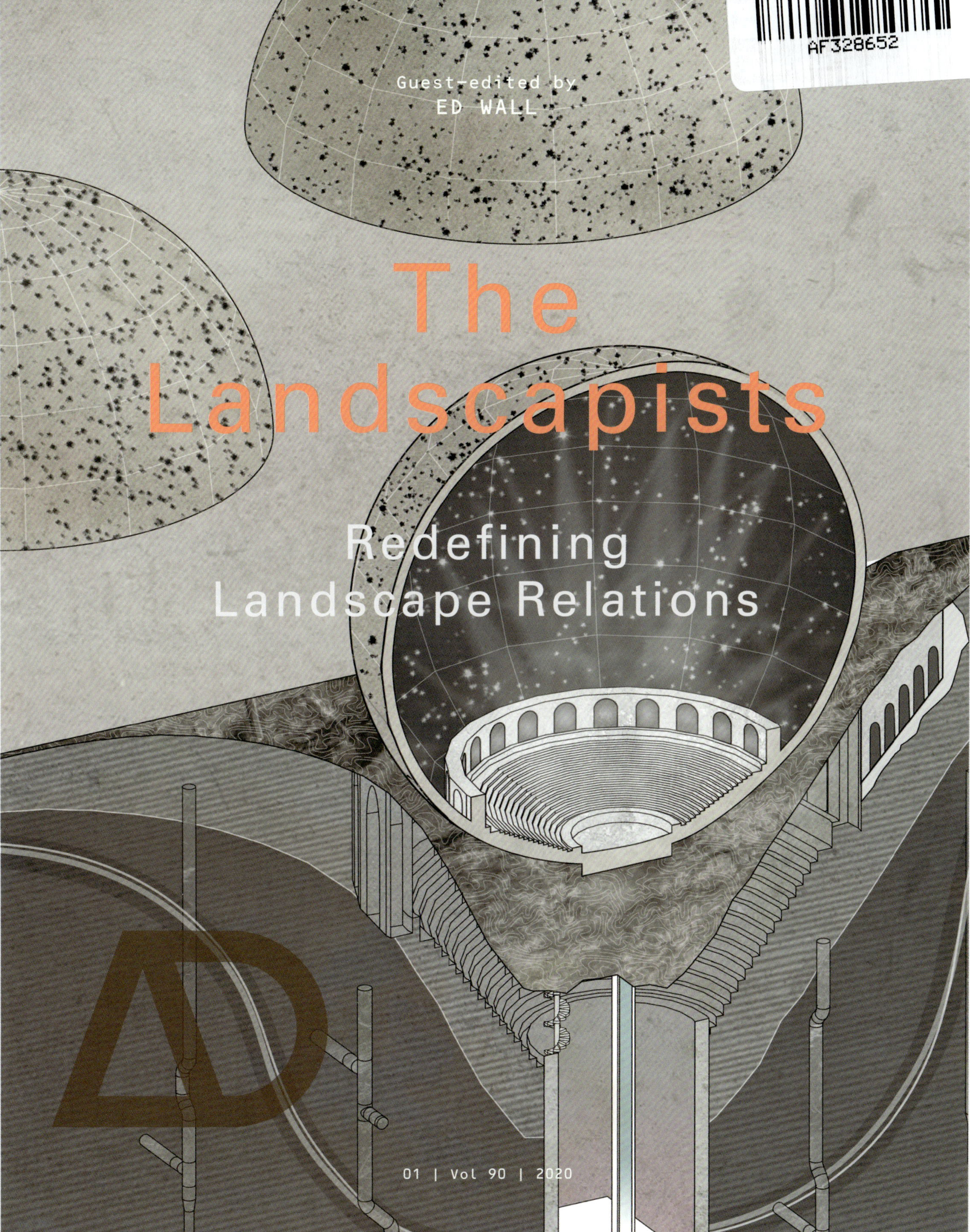

The Landscapists

Redefining Landscape Relations

ISSN 0003-8504
ISBN 978 1119 540038

Editorial Offices
John Wiley & Sons
9600 Garsington Road
Oxford
OX4 2DQ

T +44 (0)1865 776868

Editor
Neil Spiller

Commissioning Editor
Helen Castle

Managing Editor
Caroline Ellerby
Caroline Ellerby Publishing

Freelance Contributing Editor
Abigail Grater

Publisher
Paul Sayer

Art Direction + Design
CHK Design:
Christian Küsters
Barbara Nassisi

Production Editor
Elizabeth Gongde

Prepress
Artmedia, London

Printed in Italy by Printer
Trento Srl

Journal Customer Services
For ordering information, claims and any enquiry concerning your journal subscription please go to www.wileycustomerhelp.com/ask or contact your nearest office.

Americas
E: cs-journals@wiley.com
T: +1 781 388 8598 or
+1 800 835 6770 (toll free in the USA & Canada)

Europe, Middle East and Africa
E: cs-journals@wiley.com
T: +44 (0)1865 778315

Asia Pacific
E: cs-journals@wiley.com
T: +65 6511 8000

Japan (for Japanese-speaking support)
E: cs-japan@wiley.com
T: +65 6511 8010 or 005 316 50 480 (toll-free)

Visit our Online Customer Help available in 7 languages at www.wileycustomerhelp.com/ask

Print ISSN: 0003-8504
Online ISSN: 1554-2769

Prices are for six issues and include postage and handling charges. Individual-rate subscriptions must be paid by personal cheque or credit card. Individual-rate subscriptions may not be resold or used as library copies.

All prices are subject to change without notice.

Identification Statement
Periodicals Postage paid at Rahway, NJ 07065. Air freight and mailing in the USA by Mercury Media Processing, 1850 Elizabeth Avenue, Suite C, Rahway, NJ 07065, USA.

USA Postmaster
Please send address changes to *Architectural Design,* John Wiley & Sons Inc., c/o The Sheridan Press, PO Box 465, Hanover, PA 17331, USA

Subscribe to ⌂
⌂ is published bimonthly and is available to purchase on both a subscription basis and as individual volumes at the following prices.

Prices
Individual copies:
£29.99 / US$45.00
Individual issues on ⌂ App for iPad:
£9.99 / US$13.99
Mailing fees for print may apply

Annual Subscription Rates
Student: £90 / US$137 print only
Personal: £136 / US$215 print and iPad access
Institutional: £310 / US$580 print or online
Institutional: £388 / US$725 combined print and online
6-issue subscription on ⌂ App for iPad: £44.99 / US$64.99

Front cover: Richard Mosse, Koutsochero camp, Larissa, Greece, April 2016. Courtesy of the artist, Jack Shainman Gallery and carlier|gebauer. © Richard Mosse, 2016

Inside front cover: Neil Brenner and Nikos Katsikis, Map visualisation of food, feed and biofuel cropland areas, 2010. © Neil Brenner and Nikos Katsikis

Page 1: DESIGN EARTH, *Trash Peaks*, Seoul Biennale of Architecture and Urbanism, 2017. © DESIGN EARTH

01 / 2020

⌂ ARCHITECTURAL DESIGN

January/February
2020

Profile No.
263

Disclaimer
The Publisher and Editors cannot be held responsible for errors or any consequences arising from the use of information contained in this journal; the views and opinions expressed do not necessarily reflect those of the Publisher and Editors, neither does the publication of advertisements constitute any endorsement by the Publisher and Editors of the products advertised.

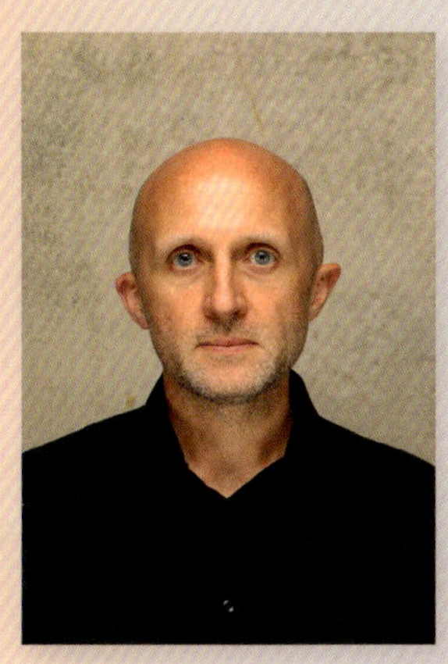

Ed Wall is Academic Portfolio Lead for Landscape Architecture and Urbanism at the University of Greenwich in London. He is also a visiting professor in the Department of Architecture and Urban Studies at the Polytechnic University of Milan, and in 2017 was City of Vienna Visiting Professor for urban culture, public space and the future – urban equity and the global agenda at the Interdisciplinary Centre for Urban Culture and Public Space (SKuOR) at the Vienna University of Technology (TU Wien). He completed a PhD in the Cities Programme at the London School of Economics (LSE), exploring relations between different ways that public spaces are made and remade in the context of London's urban development, and he trained in landscape architecture at Manchester Metropolitan University, and in urban design at the City College of New York. His work focuses on processes of landscapes and cities – with a particular emphasis on the production of collective and public spaces. Through critical research and speculative design practice he explores the uneven relations through which contemporary landscapes are constituted and the issues at stake as they are constructed.

Recent publications include *Landscape and Agency: Critical Essays* (with Tim Waterman, published by Routledge, 2017), and he is currently co-editing two further collections of essays, focusing on landscape citizenships, with colleagues at the Bartlett School of Architecture, University College London (UCL) and the University of Toronto, and unsettled urban routines, temporalities and contestations, with colleagues at TU Wien. He has also written widely, including for the *OASE Journal of Architecture*; *Landscape: The Journal of the Landscape Institute*; and *Topos*. He is co-founder and editor of *Testing-Ground: Journal of Landscapes, Cities and Territories*, a publication that brings established artists, designers and writers together with younger academics to debate contemporary landscape concerns.

Ed has co-organised several cross-institutional interdisciplinary conferences including 'Landscape and Critical Agency' at the Bartlett School of Architecture (2012); 'Unsettled: Urban Routines, Temporalities and Contestations' at TU Wien (2017); 'Design Agency within Earth Systems' at the Architectural Association (AA) in London (2018); and 'Landscape Citizenships' at Conway Hall (2018), also in London. He has also presented his research internationally, at the Radcliffe Institute for Advanced Study at Harvard University, and at Columbia University, Beijing Forestry University and the University of Edinburgh, among others.

Ed is the founder of Project Studio, a collaborative design platform for exploring processes of landscapes and cities. Experimental projects, such as Park Works, Lubricity and Roaming Forest, have been exhibited at the Van Alen Institute in New York; International Biennale of Landscape Urbanism in Bat Yam, Israel; Des Moines Art Center in Iowa; and the Building Centre, Garden Museum, Stephen Lawrence Gallery, Architecture Foundation and Royal Academy of Arts in London, and published internationally in the *Guardian*, *Architects' Journal*, *Building Design*, *Abitare* and *ArchDaily*.

INTRODUCTION

ED WALL

LES PAYSAGISTES

EXPANDING, PRODUCING, CONTESTED FIELDS OF LANDSCAPE

The landscape is never inert, people engage with it, re-work it, appropriate it and contest it.
— Barbara Bender, *Landscape: Politics and Perspectives*, 1993[1]

From the perspective of architectural design, the framing of landscapes tends towards physical geographies that have the potential to be reimagined, and visual representations that are constructed to explore and design them. Material territories are surveyed, drawn, reconsidered and reconstructed. Landscape drawings (including photography, visualisations and animations) are the means through which landscape futures are claimed and projected, and in the contemporary Western world they frequently adopt ego-centred scenographic traditions of landscape painting. This issue of △, however, presents many other relations of landscape. In the opening essay to her book *Landscape: Politics and Perspectives* (1993), the anthropologist Barbara Bender reminds us that 'Landscapes are created by people – through their experience and engagement with the world around them'.[2] This issue highlights the work of a group of geographers and artists, architects and theorists – and of course, landscape architects – and their ways of making and remaking what Bender terms 'other landscapes'.[3] These complex and often conflicting landscapes are entanglements of places, subjectivities, actions and bodies – migrant, citizen, traveller, resident and indigenous populations.

Ed Wall / Project Studio, *Drawing 7*, Valley Project, 2019

The complex of interactions across the valley are complicated by contemporary technologies, landscape processes and work practices. *Drawing 7* represents these entangled relations, including historical accounts and future imaginaries, across a valley region in the north of Scotland.

From house bands to murals and from industrial sites to imagined futures, landscapes are created, transformed and frequently contested as people situate themselves in their surroundings and in relation to each other. Practices of working, commuting, eating, drawing, imagining and dismantling, in addition to more disruptive occupations, resistances and strikes, all have the potential to produce landscapes. But while it can be argued that everyone creates landscapes, there are some individuals, communities and organisations that actively engage in relations of landscape to further its agency: exposing political agendas behind territorial claims; working with contradictions in border territories; addressing ecological challenges of climate change; exploring potentials of other worlds; critiquing processes of urbanisation; and experimenting with visual representations. These engagements are less focused on disciplinary claims to landscape as tightly defined practices of art, architecture, geography or ecology, and more part of an expanding field that is contesting, reframing and producing. Landscape is often described as a 'way of seeing',[4] as Denis Cosgrove quotes the art critic John Berger – a Western tradition of visually dominated perspectives of the land. But as this 𝝙 reveals, landscape is not just a visual medium; it can also be understood as ways of thinking, ways of working and ways of being. If we are all landscapes – material forms of daily traditions inseparable from designs for the future – then seeking to make explicit the many ways from which landscapes are constituted becomes essential.

Relations Between

To investigate relations of landscape is to question practices of how landscapes are produced, imagined, situated and lived. Patrick Geddes's *Valley Section* drawings, longitudinal sections that typically follow the course of a river from its source to the sea that were first published in 1909, remain a useful point of departure in understanding work practices and their associated tools in relation to different physical geographies. While historical accounts of Geddes's work are well documented, in particular through the research of historian Volker M Welter,[5] closer readings of his many *Valley Section* drawings in the contexts of contemporary urbanisation and an expanding frame of landscape, highlight the potential of new and critical perspectives.

In his 1909 *Valley Section*, Geddes presents the regional relations between settlements of cities and villages, productive landscapes of agricultural fields and forestry, and resources from fishing and mining. Subsequent drawings describe relations along streets, include different occupations and tools, and are situated within specific geographic contexts, such as Edinburgh. Geddes experimented over decades with different combinations, categorisations and representations. He approached his *Valley Section* drawings as Christina Leigh Geros (see her article on pp 14–21 of this issue) considers landscape architecture, as an ecology of practices.

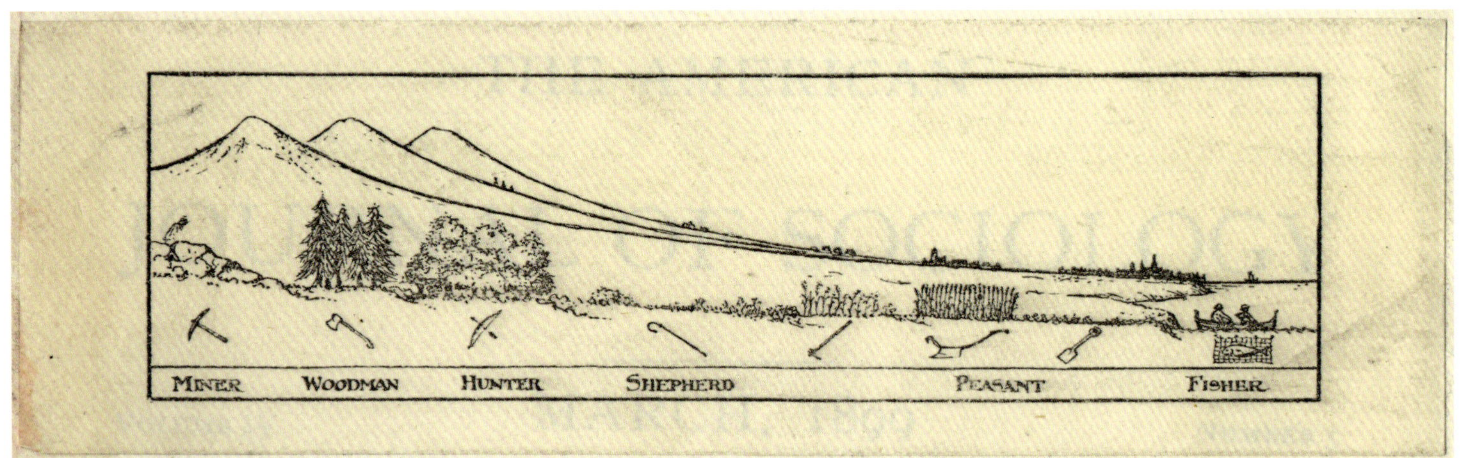

The Valley Project critically unpacks Geddes's model within the context of contemporary urbanisation to explore new landscapes, varying practices of work and their associated tools, in addition to the products of the landscape and the potential of other forms of representation. Model 1 relates the conditions of making and the tastes of whisky to a specific contemporary section across Scotland.

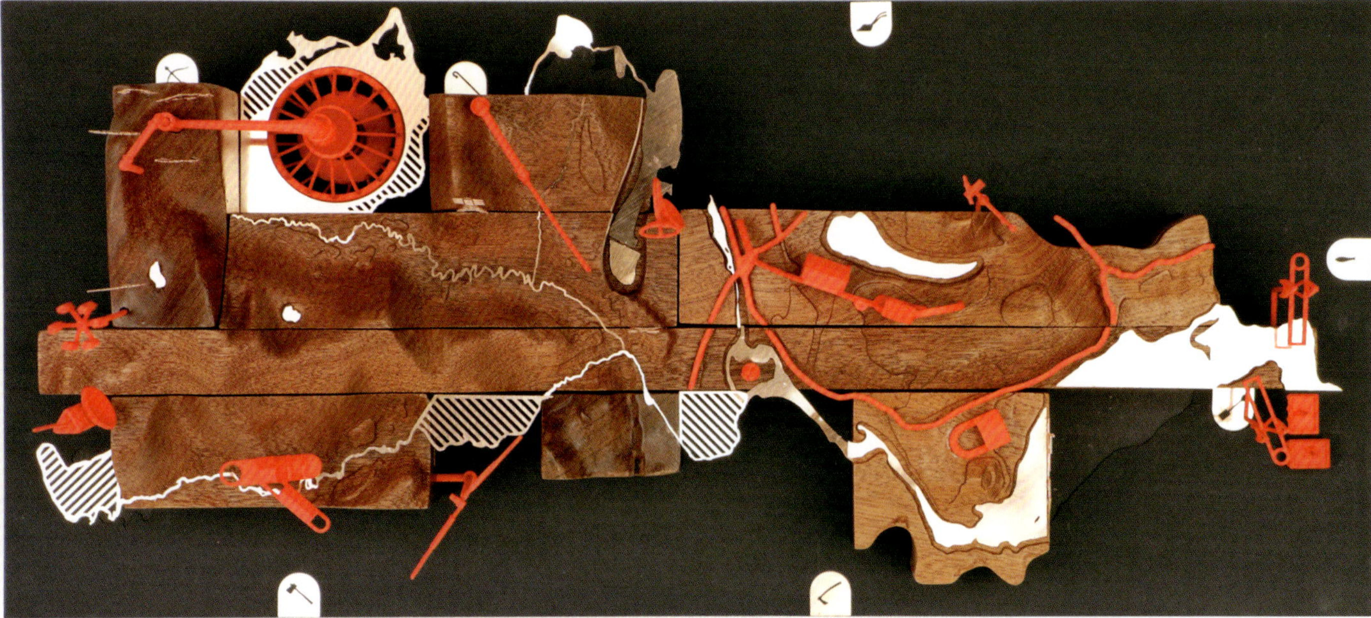

Patrick Geddes,
Valley Section,
1909

opposite top: Patrick Geddes studied botany and later applied his understanding of plants and animals to develop a regional model that he named the Valley Section. What became a series of investigations illustrates the complex relations between social occupations of humans and the environments in which these are practised.

Patrick Geddes,
Valley Section,
1925

opposite bottom: Geddes developed many different *Valley Section* drawings, that included contrasting landscape types, such as the high street, and varying occupations and tools. The illustration here shows social types in their native habitats and in their parallel urban manifestations.

The contributors to this issue seek to answer a number of recurring questions that reflect the trajectory of Geddes's investigation: What are the contemporary and future landscape relations that necessitate greater scrutiny? What are the changing techniques and emergent technologies that inform them? How can we understand these landscapes as always in process? What is at stake as they are produced? How can we work with the partial knowledge of landscapes that are constantly made and remade?

Latest Landscapes

In recent decades, changing processes of urbanisation have transformed landscapes around the world and reframed our conceptualisation of them. New infrastructures, contrasting approaches to their provision and varying patterns of growth have further complicated Geddes's regional adjacencies of mountains to forests, to fields, to cities, to oceans. As informal settlements intersect with river systems, as governments reinforce national borders, and as design practices invent new urban forms to address growing ecological crises, re-reading the physical environments of the past and questioning those of the future has become a necessary task. Tensions between landscape relations, processes of growth, decay, building, occupation, abandonment and erosion require closer critique. Such concerns are the focus of both Luis Callejas and Charlotte Hansson, who examine a series of disputed island territories (pp 46–53), and the discussion of design projects confronting climate crises described by Kate Orff (pp 94–9). The role of design is core in imagining future forms of landscapes, whether as speculative design narratives in Tiago Torres-Campos's exploration of Manhattan's geological conditions (pp 38–45) or the experimental pedagogies of Advanced Landscape and Urbanism at the University of Greenwich in London (pp 54–61).

Technological Practices

Historical work practices – miner, woodsman, hunter, shepherd, peasant, fisher – and their associated tools have also evolved. Since Geddes's investigations of the Valley Section, different technologies have transformed ways, places and routines of work. Flexible working patterns and digital devices that uncouple social activities from specific landscape types, and blur online and physical realms, have become new landscape tools (see Tim Waterman's article on pp 80–87). Machines for observing and recording, satellites communicating from earth's orbit, and the means to survey street spaces and activities also continue to be advanced. Contrasting ways of reading landscapes are the focus of Richard Mosse's photography, which, as on the cover of this 𝐷, highlights the lived experience of refugees through the use of a thermal video camera. Different methods are also key to collaborations across disciplines, such as the aerial photography of Alex McLean and corresponding collages constructed by James Corner, who read the Jefferson grid from the road. The resultant publication, *Taking Measures Across the American Landscape* (1996),[6] provides a point of reflection for Corner in his description in this issue of different measures of contemporary urban landscapes (pp 88–93). Infrastructural forms and technologies are also the focus of DESIGN EARTH (Rania Ghosn and El Hadi Jazairy) who visualise narratives of waste infrastructures through their *Trash Peaks* project (pp 32–7) to create new understandings and approaches to landscape.

In Process

Growth, decay, production and waste are bound up with all landscapes. As Bender explains, and as illustrated by Archigram in their section drawings of *Instant City* (1970), they are always 'in process of construction and reconstruction'.[7] In the context of less tangible relations between people and their surroundings, and slower-moving

H5,
My Generation,
2018

My Generation is a short animated film by H5, a creative studio based in Paris. Building on the acclaim of their earlier short film *Logorama*, an animation that critiqued consumer culture and its impact on global warming, in *My Generation* H5 address landscapes of finance, politics, sex, religion, sport, data and art culture in the context of growing populist rhetoric.

changes such as localised weathering and impacts of global warming, as investigated, for example, by Studio Folder in the Italian Alps, landscapes can seem frustratingly illusive and impossible to represent. However, the geographer Doreen Massey reminds us that in addition to a conceptual understanding of landscapes as processes, we also encounter them as material objects: 'Of course, in the practical conduct of the world we do encounter "entities", there *is* on occasion harmony and balance; there *are* (temporary) stabilisations; there *are* territories and borders'.[8] Registering the physicality of landscapes in the context of the nonhuman processes from which they are produced, and human practices by which they are constituted, brings complex considerations of landscape into close proximity. The quarry situated in the mountains, the minerals extracted, the individual workers, the corporate owners and the tools employed are inextricably connected with planned factory towns, pollution of local rivers, commodities consumed and landscapes of waste and reuse. Such messy associations, further complicated by our individual and collective experiences, define all landscapes, as Harry Bix describes of East Anglia Records' interconnectedness of quotidian musical performances, drawings and poetry (pp 62–5), and as Gareth Doherty and Pol Fité Matamoros (pp 100–105) and Teddy Cruz and Fonna Forman (pp 114–19) illustrate in their research of border landscapes.

Uneven Sections

The equal division of the Geddes *Valley Sections* across landscape types and their associated occupations fails to represent uneven processes of production that result

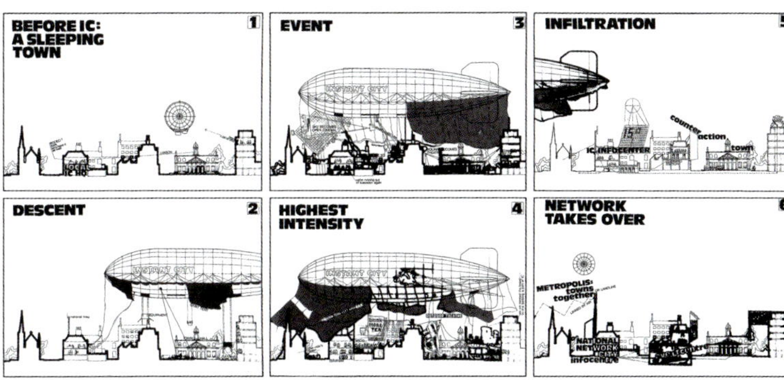

Peter Cook, *Instant City Airships, sequence of effect on a typical English town,* 1970

This series of sectional drawings of *Instant City* emphasises both the temporality of the proposal as well as its longer-term impact on the potential to make. Comprising many drawings and models by the avant-garde architecture group Archigram, the project works across a range of scales demonstrating site-specific considerations from architectural forms to regional- and national-scale networks.

Valentina Galiulo,
Protocol map of Environmental Justice (Seville),
TELLme project,
Department of Architecture and Urban Studies,
Polytechnic University of Milan,
2019

The complexity of creating mapping systems that are sufficiently adaptable for different geographic contexts while adequately consistent to provide for comparison is at the core of the EU-funded TELLme project led by Professor Antonella Contin at the Polytechnic University of Milan. In this specific map, the concept of environmental justice is combined with social cohesion, referring to the people who inhabit urban space and the relationships established between individuals and the territory.

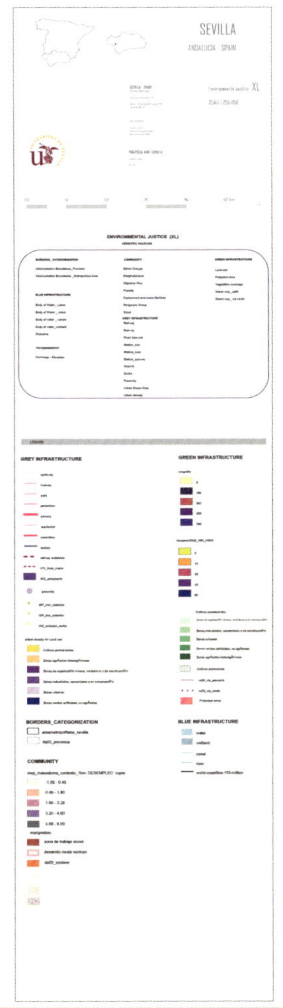

in intensities of investment in some areas and lack of infrastructure in others. Furthermore, the absence of certain work activities and the invisibility of landowners, planners, developers and politicians, individuals and organisations that make political decisions and inform economic agendas, ignore the imbalances of power across contemporary landscapes. While the sectional projections employed in Geddes's drawings offer a relational understanding of the region, their diagrammatic qualities also deny the subtleties and complexities of landscapes in change. Neither do they represent the unequal distribution of risks endured by some people and the benefits gained by others. In this issue, Pierre Bélanger (pp 120–27) identifies conflict between indigenous cultures and Western practices, as techniques and languages are simultaneously appropriated and denied their historical significance, while Matthew Gandy (pp 106–13) identifies subtle contradictions within urbanisation practices in his research into Park am Gleisdreieck in Berlin that may suggest more significant intertwinings of urban change. The specificity of processes of change that unfold in different places is highlighted in Toya Peal's description of the site-specific works of Berlin-based artist Larissa Fassler (pp 66–73) who aims in her work to represent the lives and concerns of people inhabiting the city.

Incomplete Landscapes
In 'Situated Knowledges: The Science Question in Feminism and the Privilege of Partial Perspective' (1988), Donna Haraway states that the 'view of infinite vision', provided by the technological enhancements of human sight through surveillance systems, video displays, graphic manipulation and mapping, 'are an illusion'.[9] While not explicitly claimed by Geddes, his *Valley Section* drawings imply a distanced, impartial and total view of landscapes. By recognising the selective and limited vision provided by satellites, cameras and surveying equipment along with the partial understanding of ourselves and limited knowledge of our environments, we can appreciate an incompleteness in all landscapes. Totalising vision can mistakenly suggest a comprehensiveness of order and control of what makes up the relations of geographies, practices and tools.

The suggestion of the valley as a contained system also denies wider global relations that are essential to practices addressing ecological destruction and anthropogenic climate change. Reconsiderations of the Valley Section, as explored in the recent work of Project Studio – a platform for design and research collaborations focused on landscapes and cities – must address the relationships between planetary scales of urbanisation and other lives, spaces and processes, including what Neil Brenner and Nikos Katsikis term 'operational landscapes' (pp 22–31). Haraway states: 'The moral is simple: only partial perspective promises objective vision'.[10] She advocates adopting viewpoints from below, from positions of the subjugated: 'there is good reason to believe vision is better from below the brilliant space platforms of the powerful'.[11] However, mediating between perspectives, interrogating the structures of power as they manifest in specific situations, provides a more focused approach. To understand the relations between landscapes of geopolitical decisions and global economic agendas and how they come

Drawing Architecture Studio,
Tuan Jie Hu Panorama,
Beijing,
2014

Vividly depicting the daily life of a neighbourhood through exploring different drawing techniques, the sectional projections of buildings combined with the opportunity for the viewer to simultaneously read different projections challenges conventional representational methods and offers a particular relational perspective of this urban landscape. The drawing describes the complexity of relations between people living their lives, the urban environment in which they live, and the audience.

Studio Folder,
Italian Limes,
2019

Italian Limes is a research project led by Studio Folder (Marco Ferrari, Elisa Pasqual, Alessandro Busi and Aaron Gillett) and an interactive art installation that explores remote Alpine regions, where national borders drift as glaciers move and change. The researchers monitor Austrian-Italian watersheds to look at relationships between borders and environmental change.

to bear on specific places, lives and practices requires a range of strategies, representations and actions – as seen in the environmental protests of Extinction Rebellion. The consideration of multiple, less fixed positions from which different and collective approaches can be considered are key in challenging the ego-centred landscapes that Bender critiques.[12]

The Landscapists

Who and what define landscapes? Which practices are employed as landscapes are formed? What is at stake in their production? The landscapes of this issue of *D* are contrasting unstable urban ecologies, creative environments and contested territories, connected through reflecting and situating the processes and practices from which they are constituted. As Bender describes: 'Each individual holds many landscapes in tension'.[13] This *D* exposes such conflicts, evidencing divergent ideas, provoking contradictions and forming moments of connection within and between. Critical urban and landscape theory can be uncomfortable bedfellows with contemporary design practices, as the role that architectural practice has in contributing to gentrification and ecological damage is questioned or as theoretical writing is tested by the pragmatics of client briefs, financial ambitions and site constraints. Furthermore, landscape does not sit neatly in a line-up of disciplinary silos. As Bender insists, landscape 'has to be an area of study that blows apart the conventional boundaries between disciplines'.[14] This need to continually unpack, reconsider and reconstitute the professional frame of working with landscape is highlighted by Charles Waldheim as he recalls historical struggles for the naming of the disciplines.[15] He describes the French term *'paysagiste'* that has been used to refer to those undertaking a range of landscape practices. However, he explains that in contemporary Paris landscape architects have reappropriated the word as less formal than the conventional term for landscape architect, *'architecte-paysagiste'*: 'The translation of the term into English,' he explains, 'offers itself readily as simply "landscapist"'.[16]

Extinction Rebellion protest,
Waterloo Bridge,
London,
18 April 2019

The week-long occupation of London's Waterloo Bridge by the environmental activist group Extinction Rebellion created a new landscape in the capital in order to provoke political and economic action to address the ongoing ecological and climate crisis. Increasing awareness of the relation between human actions and planetary ecologies establishes specific global-scale landscapes.

Notes

1. Barbara Bender, *Landscape: Politics and Perspectives*, Berg (Oxford), 1993, p 3.
2. *Ibid*, p 1.
3. *Ibid*, p 2.
4. See Denis Cosgrove, 'Prospect, Perspective and the Evolution of the Landscape Idea', Transactions of the Institute of British Geographers, 10 (1), 1985, pp 45–62, and John Berger, *Ways of Seeing*, Penguin (London), 1972.
5. Volker M Welter, *Biopolis: Patrick Geddes and the City of Life*, MIT Press (Cambridge, MA), 2002.
6. James Corner and Alex McLean, *Taking Measures Across the American Landscape*, Yale University Press (New Haven, CT). 1996.
7. Bender, *op cit*, p 3.
8. Doreen Massey, 'Landscape as a Provocation: Reflections on Moving Mountains', *Journal of Material Culture*, 11 (1/2), 2006, p 40.
9. Donna Haraway, 'Situated Knowledges: The Science Question in Feminism and the Privilege of Partial Perspective', *Feminist Studies*, 14 (3), Autumn 1988, p 582.
10. *Ibid*, p 583.
11. *Ibid*.
12. See Ed Wall, 'Post-landscape or the Potential of Other Relations with the Land', in Ed Wall and Tim Waterman (eds), *Landscape and Agency: Critical Essays*, Routledge (Abingdon), 2017, pp 144–163.
13. Bender, *op cit*, p 2.
14. *Ibid*, p 3.
15. Charles Waldheim, *Landscape as Urbanism: A General Theory*, Princeton University Press (Princeton, NJ), 2016, pp 171.
16. *Ibid*.

Christina Leigh Geros

DESIG
MOME

SITE,
MEDIA AS

NING NTUMS

Christina Leigh Geros,
The Global Monsoon,
Monsoon Assemblages,
University of Westminster,
London,
2019

An experimental exploration of the cartographically constructed world through the annual cycles of the monsoon's atmospheric centres of action, produced through data and research of the monsoon that has been compiled since 1875. In its action, the monsoon brings only certain corners of the globe into focus.

PRACTICE, LANDSCAPE

Christina Leigh Geros believes that landscape is more than the land, and that landscape architecture is a process of designing and reconciling disparate 'momentums'. A tutor in the MA Environmental Architecture programme at the Royal College of Art, research fellow for Monsoon Assemblages at the University of Westminster, and design research strategist for PetaBencana.id, here she explores the expanding of the definition of 'landscape' through wider analysis of contemporary media, site and practice.

To be concerned with and about the environment is to imagine a practice of being and a way of seeing, inhabiting and designing momentums. Landscape architecture is well positioned to explore environmental concerns by de-centering the human through an expanded acknowledgment of shifting relationships to others and their transformations. In the context of two current design-led research projects – The Orang-orang and the Hutan, within the Royal College of Art's MA Environmental Architecture Programme; and Monsoon Assemblages, led by Lindsay Bremner at the University of Westminster, both in London – the place of the designer and the moment of design are reimagined in the search for knowledge, its formation and infrastructure.[1] At the core of both projects, landscape architecture is considered as an ecology of practice which aims to address questions of who is designing, with whom and to what end. Each project begins from an understanding that 'landscape' is no longer a-thing-out-of, nor is it a-space-within our built environments; rather, *we* are landscape – our thoughts, actions and inactions enmeshed within ecologies. Engaged in both 'field' and archival research, each project aims to build itself from within conversations, drawing out questions that offer opportunities for realignment and growth able to produce material and political change capable of mobilising energies, human and nonhuman, into action.

To begin, it is necessary to provide some contextual framework for the practice of landscape architecture, as both projects are multidisciplinary. To accommodate the identification of 'landscape as site' and the emergence of 'landscape as practice' within different conceptual frameworks, both projects acknowledge landscape as both a territory of human–ecological entanglements and as a means to uncover and engage with sources of degradation and renewal. Additionally, both projects are long-term applied research endeavours and, as such, are inprogress – continually redefining both objectives and methods as they mediate between 'field' and 'archive'.

Landscape as Site

Borneo, in particular the carbon-rich soils that construct the biologically diverse ecosystems of the tropical peatland, as a site of investigation and intervention requires human activity to be addressed within geological processes of accumulation and extraction. The natural structure of the peat bog provides deep pockets of oxygenated organic matter that produces long, smouldering fires. These fires restructure soil nutrients and water-holding capacities into more stable matter capable of a full index of fertility. Over thousands of years these assemblages of human–forest relations, or anthrosols, have transformed pockets of nutrient-poor soils into highly productive lands. The 'site' of The Orang-orang and the Hutan transcends the surface, engaging with micro-characteristics produced by and productive of human activity.

An architect of time and space around the globe, the monsoon constructs ecologies, customs and cultures – environments – within and across landscapes. Itself an assembling of earth-system dynamics, it is experienced differently, at different scales, around the globe; yet has become synonymous with South Asia. Focused around the Bay of Bengal – from Chennai, India to Dhaka, Bangladesh and Yangon, Myanmar – Monsoon Assemblages engages with the 'site' of the monsoon as both a material and cultural construct embedded within social environments, both human and nonhuman.

Acknowledging that landscape is more than land requires a shift in focus. The terms 'land' and 'landscape' carry connotations of the natural, the aesthetic and the agricultural along with histories of legal and territorial ideas of national politics and cultural identities. As Kenneth Olwig suggests, 'customs and culture defined a Land, not physical geographical characteristics';[2] thus, landscape is not bound to land, but is held within an environment.

Landscape as Practice

A situated understanding of landscape as a moment within and amongst environments – a constellation of relational consistencies of coexistence – bears in mind that any landscape must be horizenless, ever evolving, and always becoming. Approaching this as a site of investigation necessitates an expanding field of practices which can be utilised within landscape architecture and considered 'a tool for thinking through what is happening'.[3] For centuries, the spatial arts have operated as interrelated ways of articulating, sharing and shaping culture[4] – performing as non-neutral tools for seeing and making. Working across disciplines and within ever-evolving 'sites', the interventive nature of design often finds itself in the articulation of questions which may mobilise momentums towards previously unimagined environments.

opposite: Working with satellite imagery and GIS software, MA student Yu experimented with colour band combinations (6, 5 and 2) to expose patterns of active burn and historical char in the detection of fire activity in peatlands.

Monsoon Assemblages is a five-year research project funded by the European Research Council (ERC) under the European Union's Horizon 2020 research and innovation programme (Grant Agreement No. 679873) and begun in 2016. While multidisciplinary, it is a design-led project that focuses a critical lens on established and changing relationships between the people, lands and cities of the monsoon through novel approaches to materialisms, cartographies and narratives. In the 'field', photography, film, drawing and interviews are used to uncover ways of knowing and living with the monsoon that may elude mainstream narratives. Often these techniques capture large- and small-scale narratives that present challenges inherent to processes of combining qualitative and quantitative data of different timescales and mediums – yet are essential to drawing out changing relations to staid notions of climate. For example, with the Bede – or 'river gypsies' – of Louhajong, Bangladesh, conversations with individuals quickly grew to large, animated meetings with the community and a drawing of a 'monsoon calendar' constructed around patterns of seasonal movement timed with river allowances afforded by monsoonal flows from the Tibetan Plateau. Upon return to London, these accounts try to find resolution within the 'archive' of monsoonal time and space.

In the 'field', Beth Cullen anthropologist for Monsoon Assemblages, draws a version of a monsoon calendar as it emerges from conversations with the Bede community; while Christina Leigh Geros documents the process with video and photographic footage.

The Orang-orang and the Hutan began in 2019 as a four-year studio research project. Framed within the academic studio, it engages design students with local environmental activists, geographers, farmers, filmmakers and researchers in Borneo to address trans-scalar concerns, from the soils that bear them to the airs that carry them. Born from extractive agricultural practices, conflicts over indigenous lands and rights of inhabitation converge through knowledge and production of soils. In the 'field', students work in collaboration with residents – of different genders, ages, occupations and land tenures – to uncover knowledge about these soils, passed from person to person, generation to generation, that may offer material proof of occupation. On a recent trip to Mantangai in Central Kalimantan, residents and students used drones to follow a centuries-old boundary between one village and another. Once marked by the periodic placement of vertical posts, the acidic waters of the peat swamp had dissolved evidence of the path's intentional placement; yet seen from the air, the path's dimensions and geometry – a clear departure from its surroundings – could clearly be traced. This boundary was not a marker of property, but of soil knowledge – each side of the line productive of a particular crop – proof of cohabitation and soil production long preceding the government's claim to land. In the studio, students continue to collaborate with their co-conspirators in Borneo to find ways to map this knowledge and construct an authoritative 'archive' of the unseen and erased.

In the 'field' of Central Kalimantan, students explored the canals of Borneo's peatlands with environmental activists from Indian NGO WALHI and local residents, in search of methods of representation of nearly lost historical boundaries.

Yu worked with local residents and environmental activists from WALHI in Central Kalimantan to begin mapping 'hidden' markers of indigenous land tracts using aerial footage enabled by drone imagery.

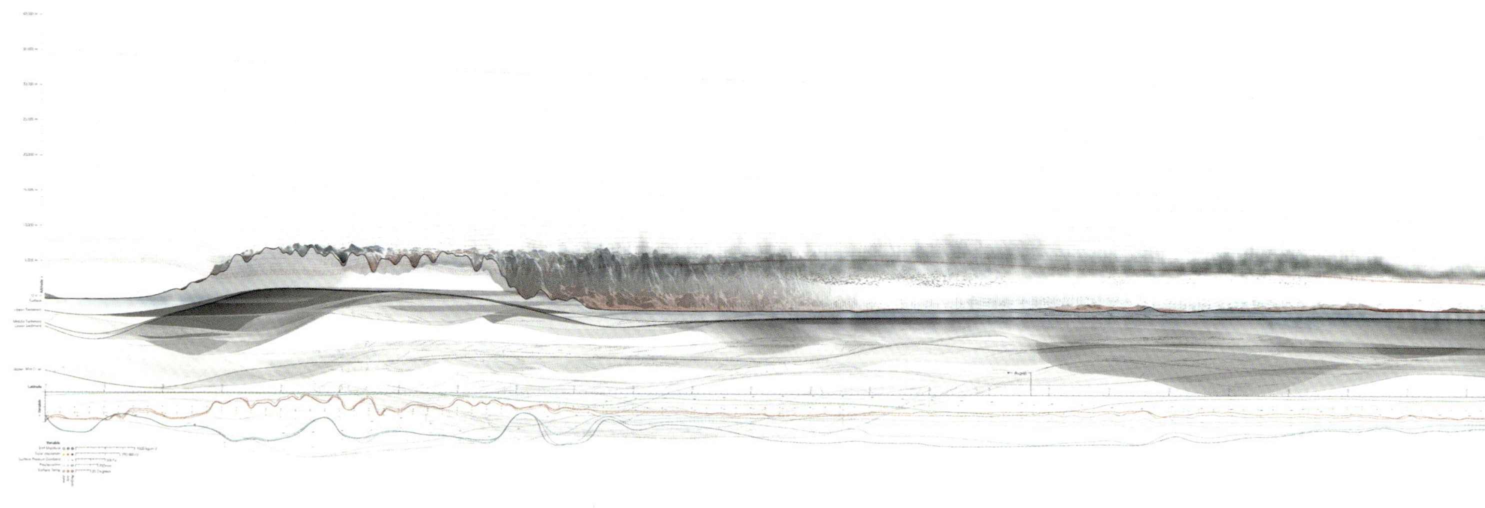

Ecologies of Media

Expanding landscape 'site' and 'practice' also questions the medium and collaboration of design. Monsoon Assemblages and The Orang-orang and the Hutan both propose 'sites' that are cultural and institutional constructions of social and scientific media. Both projects aim to be interventive and impactful, while driven by process and not geared towards solutionist propositions and final outputs; but what does that mean for design? Towards design imaginaries for projects that are in progress, it is useful to engage with another project that probes 'site' and 'practice', allowing a process of engagement to perform as 'media' and 'collaboration' that is generative towards landscape and design.

Lying within monsoonal territory and home to more than 30 million residents, built atop a deltaic plain traversed by 13 rivers, Jakarta, Indonesia is affected by regular and significant flooding. However, in a tangle of concrete roadways and high-rise buildings, the river network – the actual pumping heart of the city – is often obscured from view and unable to mediate the city's inundation. Working across disciplines including design, geography, computer science and philosophy, and employing a diverse set of methods for research, PetaBencana.id has created an online monitoring and coordination platform that allows residents to help one another during flood events.[5] Based on ethnographic research, the design of the platform learns from the residents of the city and expands the already existing epistemological network of river knowledge into an online geospatial conversation –

publicly built and accessible – drawing the river back into the city's consciousness. Inhabiting the riverbanks and bearing the highest levels of risk, the city's urban poor acknowledge a 'belonging' with the river; while the city's more affluent residents often neglect to recognise themselves as sharing the same urban ecology. In mapping conversations about flooding, the resultant image draws the river and the city together – into shared belonging and shared care.

Like PetaBencana.id, Monsoon Assemblages and The Orang-orang and the Hutan work between the 'field' and the 'archive' to create media that enable a continued learning process – a collaboration – with their environments. Beyond illustrative imaging, these new constructions of knowledge aim to be instructive of new questions, new engagements and new momentums.

Archival research of the monsoon reveals an evolution of cartographic constructions since the 17th century. Processes of 'unfolding potential', these cartographies 'enable, rather than depict' the spaces and relationships of the monsoon through particular moments of social and institutional construction.[6] Most often constructed as revenue-building infrastructure or weather-induced risk, today's changing monsoon now asks to be reconceived of as landscape – a moment of mediation between two horizons and a site within which the built environment is entangled. From policy to design, the enabled imaginary of a kinetic site is a question of landscape architecture at the very edges of its practice. How might the world appear through the lens of the monsoon?

Lying within monsoonal territory and home to more than 30 million residents, built atop a deltaic plain traversed by 13 rivers, Jakarta, Indonesia is affected by regular and significant flooding

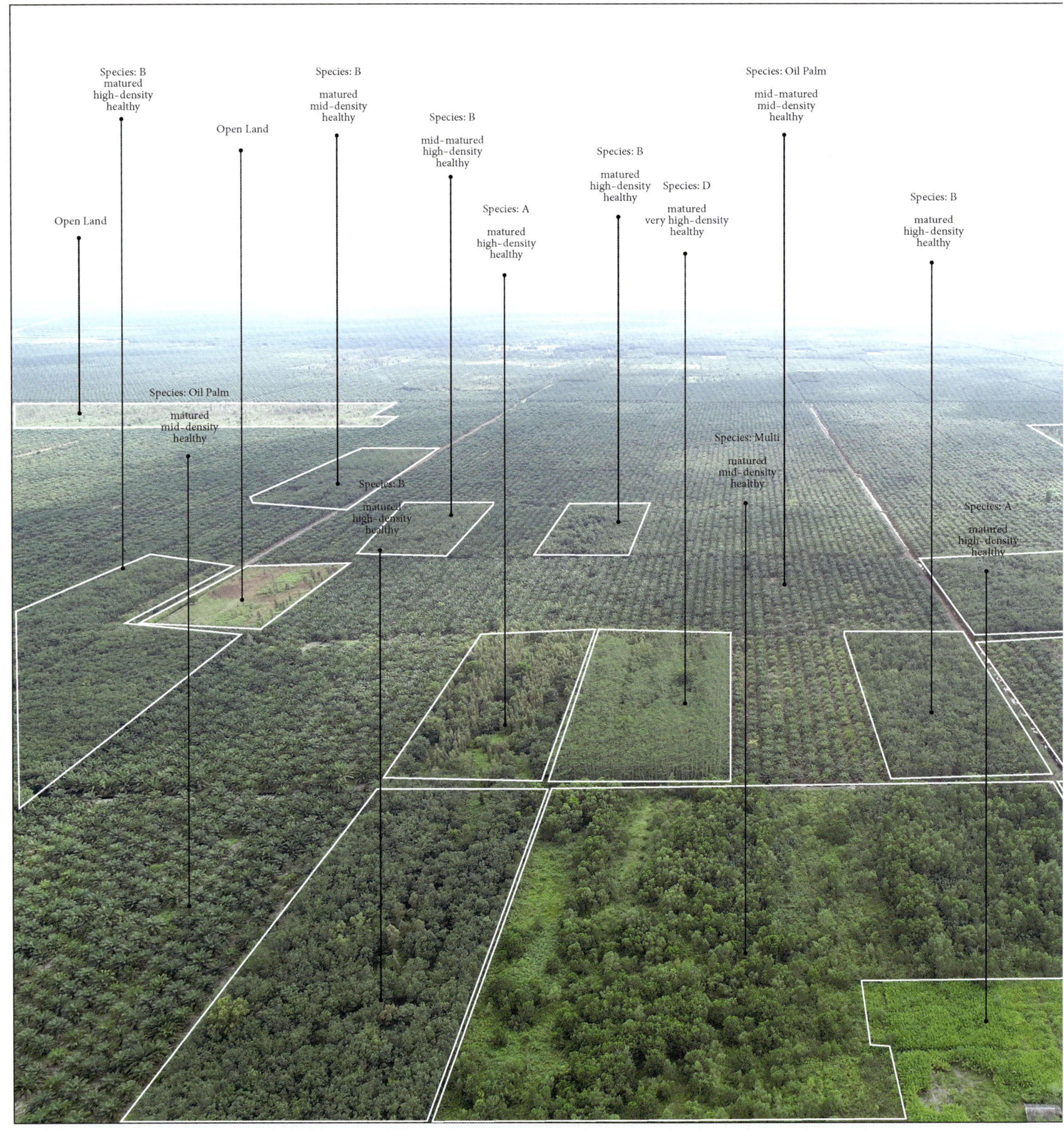

From hyper-micro to extra-macro, the human and nonhuman actors constituent within the local and global reach of this environment must be considered as designers of a landscape

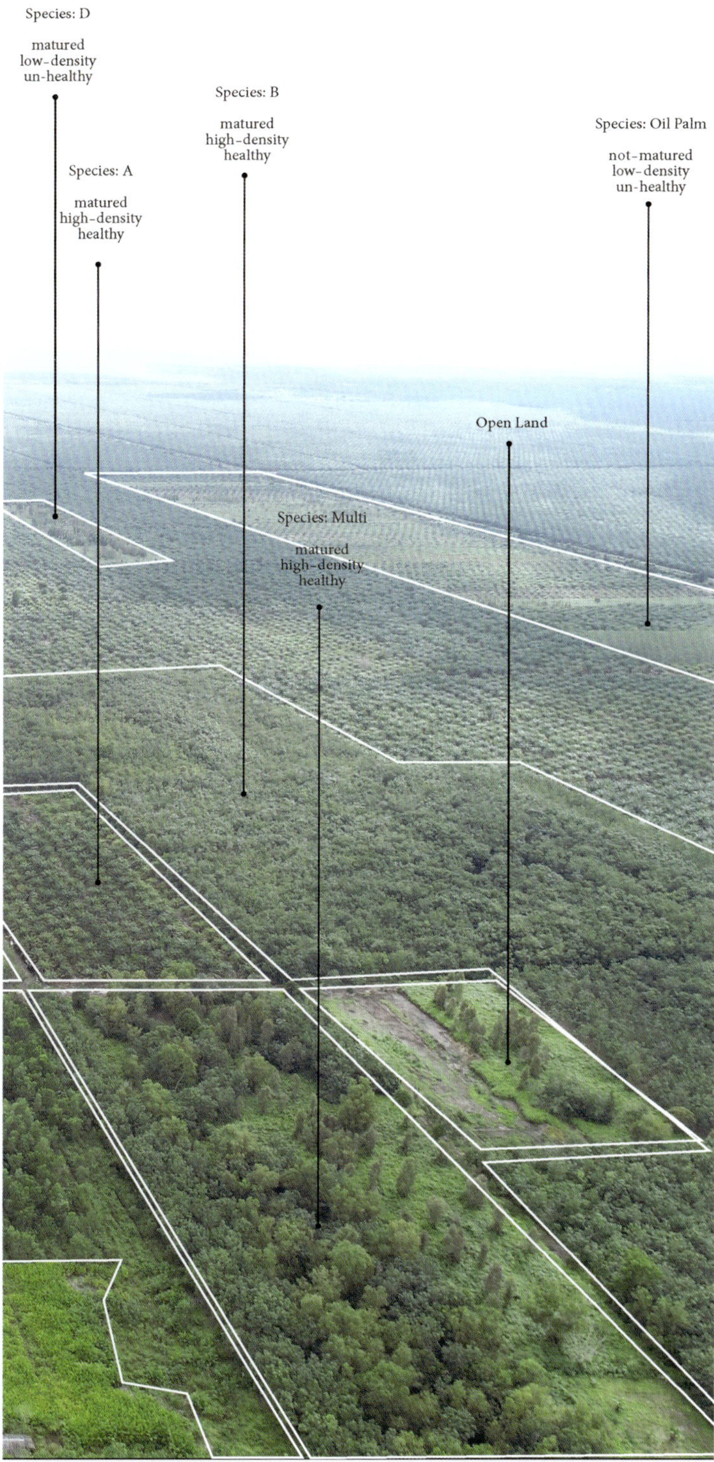

Kaiwen Yu,
Discrepant Cartographies,
The Orang-orang and the Hutan,
MA Environmental Architecture,
Royal College of Art (RCA),
London,
2019

Yu begins his investigation into the 'discrepant cartographies' of Borneo's peatlands – asking what can be seen, from what distance, with what 'eye', and what can be known from it?

By contrast, the soils of the Bornean peatland lack representation and their *terra-nullius* status constructs their vulnerability. Exposing the epistemological frameworks that have produced these soils through centuries of cohabitation may enable these soils to voice themselves within their environments. Detailed knowledge of specific plant species, the colour of the smoke released from burning grounds and the response of soil to the sole of a boot can build an atlas of knowledge about the horizontal and vertical dimensions of a peat bog. The generation of media to collect, collate and represent this knowledge has the potential to intercede within the current crisis produced by the industrial commodification of under-represented soils. From hyper-micro to extra-macro, the human and nonhuman actors constituent within the local and global reach of this environment must be considered as designers of a landscape. In the performance of due diligence, shouldn't one ask: What is the cost of one cubic hectare of carbon? And who will pay the price?

Media of Landscape

Whether representing existing epistemologies or attempting to weave themselves into these networks, both projects aim to produce new ontologies within shared neuro-ecological landscapes. As each participates in a hyper-sensed and recorded world, a continual process of engagement allows opportunities of material, political and social change to emerge through interactive landscapes. An active evolution of landscape architecture – always at the edge of bounded discipline – allows it to attend to shifting environments; while designing shifts through and towards engaged landscapes that place the environment as the object of concern and cooperation. As long-term engagements of interventive experiment addressing questions of environmental representation, these projects operate as disciplinary probes – processes of questioning and redefining partnerships and methods of design action. ⌀

Notes
1. Christina Leigh Geros, *Studio 2: The Orang-orang and the Hutan*, Royal College of Art: https://www.rca.ac.uk/schools/school-of-architecture/environmental-architecture/studio-descriptors-201819/studio-2-orang-orang-and-hutan/; Lindsay Bremner, The Research Project, Monsoon Assemblages: http://www.monass.org/project/.
2. Kenneth Olwig, *Landscape, Nature, and the Body Politic: From Britain's Renaissance to America's New World*, University of Wisconsin Press, Madison (WI), 2002, p 19.
3. Isabelle Stengers, 'Introductory Notes on an Ecology of Practices', *Cultural Studies Review*, 11 (1), 2005.
4. David Leatherbarrow, 'Is Landscape Architecture?', in Gareth Doherty and Charles Waldheim (eds), *Is Landscape … ? Essays on the Identity of Landscape*, Routledge (Oxford and New York), 2016, p 641.
5. Yayasan Peta Bencana, PetaBencana.id, https://info.petabencana.id/about/.
6. James Corner, 'The Agency of Mapping', in Denis Cosgrove (ed), *Mappings*, Reaktion Books (London), 1999, pp 231–52.

Text © 2020 John Wiley & Sons Ltd. Images: pp 14-15 Image Christina Leigh Geros / Monsoon Assemblages. Monsoon Assemblages is a project funded by the European Research Council (ERC) under the European Union's Horizon 2020 research and innovation programme. Grant Agreement No.679873; pp 16, 18(c), 20-21 Images Kaiwen Yu; p 17 Photo Christina Leigh Geros / Monsoon Assemblages; p 18(t) © Christina Leigh Geros; pp 18-19(b) Image John Cook / Monsoon Assemblages

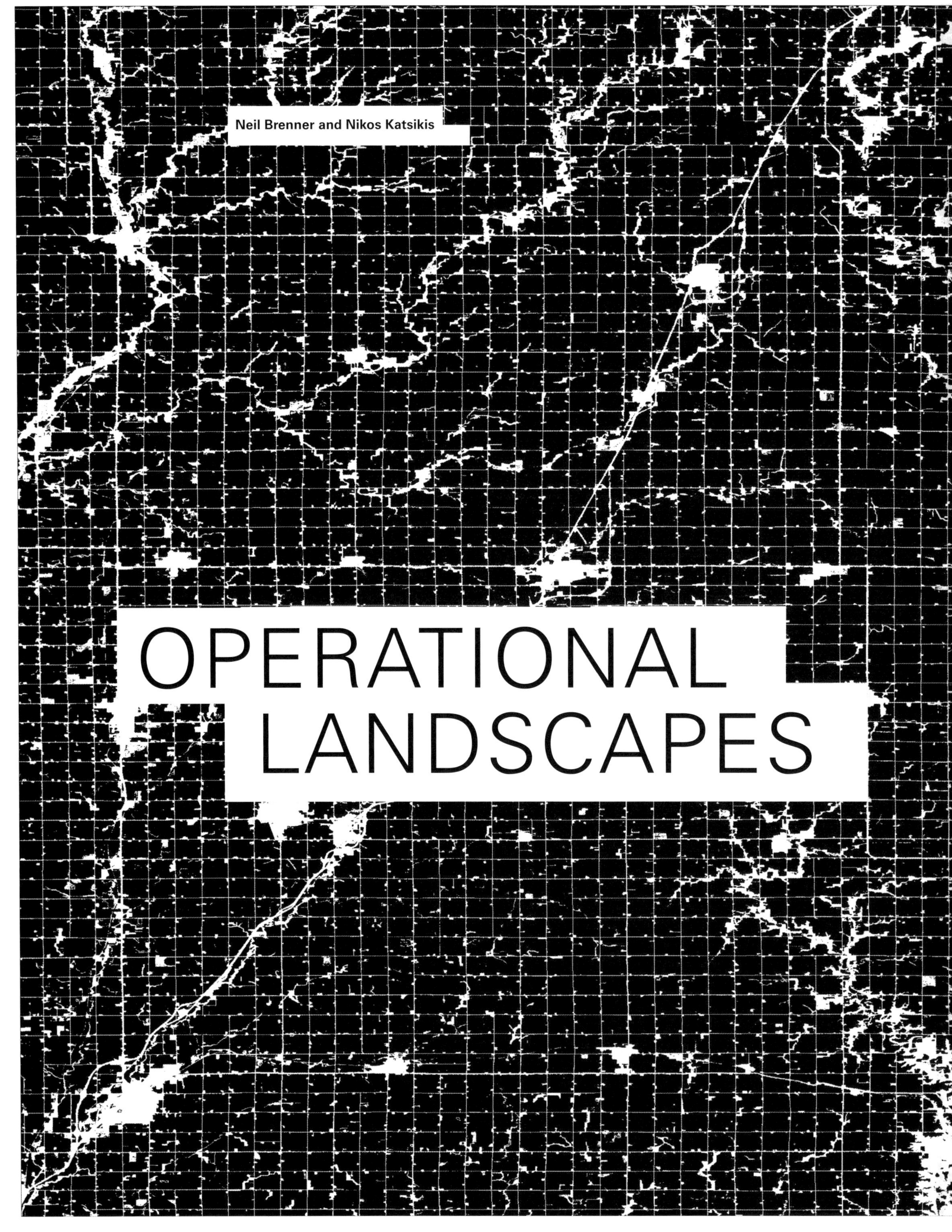

OPERATIONAL LANDSCAPES

Neil Brenner and Nikos Katsikis

Neil Brenner and Nikos Katsikis, Map visualisation of the US Corn Belt, 2018

Capital-intensive, highly industrialised and densely equipped landscapes of cash-crop monocultures dominate the Corn Belt, where more than 80 per cent of all land (depicted in black) is dedicated to the cultivation of corn and soya beans. The zone is configured among 1-mile (1.6-km) tiles within a Jeffersonian grid pattern. This permits the maximally efficient operation of agro-industrial machinery. Beneath this terrestrial surface is an extensive subterranean drainage system that supports soil tilling. Data source: USDA National Agricultural Statistics Service Cropland Data Layer (2018), published crop-specific data layer, available at https://nassgeodata.gmu.edu/CropScape/.

In recent decades, the field of urban studies has neglected the question of the hinterland: the city's complex, changing relations to the diverse non-city landscapes that support urban life. **Neil Brenner and Nikos Katsikis** of the Urban Theory Lab at the Harvard Graduate School of Design argue that this 'hinterland question' remains essential, but must also be radically reimagined under contemporary conditions.

What role do spaces beyond the city play in urbanisation, and how are they transformed through this process? City-building is a process of sociospatial concentration, but its preconditions and consequences are not confined to the city's immediate environs. The term 'hinterland' is used here to demarcate the variegated non-city spaces that are swept into the maelstrom of urbanisation, whether as supply zones, impact zones, sacrifice zones, logistics corridors or otherwise. Such spaces include diverse types of settlements (towns, villages, hamlets), land-use configurations (industrial, agrarian, extractive, energetic, logistical) and ecologies (terrestrial, oceanic, subterranean, atmospheric). We refer to explorations of such spaces, and their role in urbanisation processes, as engagements with 'the hinterland question'. Across the urban social sciences and design disciplines, the hinterland question is today considered secondary or even irrelevant to the study of urbanisation; the city, its dense socioeconomic networks and its powerful agglomeration economies occupy centre stage. In the age of planetary urbanisation, this position is untenable: city/hinterland relations lie at the heart of the contemporary urban problematique. And yet, these relations are today undergoing mutations that necessitate not only a repositioning of the hinterland question into the core of urban research and practice, but its radical reconceptualisation.

Cities Without Hinterlands?

Prior to the 1970s, the field of urban studies devoted extensive attention to the role of non-city landscapes in the urbanisation process. From Johann Heinrich von Thünen's early 19th-century model of the relationship between an isolated city and land-use differentiation in its agrarian hinterland, through the early 20th-century writings of Patrick Geddes, Lewis Mumford and Benton MacKaye on ecological regionalism, up through post-Second World War explorations of central place hierarchies and polarised regional development, city/hinterland relations were widely regarded as constitutive dimensions of the urban problematique.[1]

Few images have had a greater impact on contemporary metanarratives of global urbanisation than the 'nighttime lights of the world' series, initially synthesised during the 1990s in the National Geophysical Data Center (NGDC) in Boulder, Colorado, and subsequently improved through NASA's remote sensing networks. Data source: VIIRS DNB Nighttime Lights Composites, NOAA National Center for Environmental Information (NCEI).

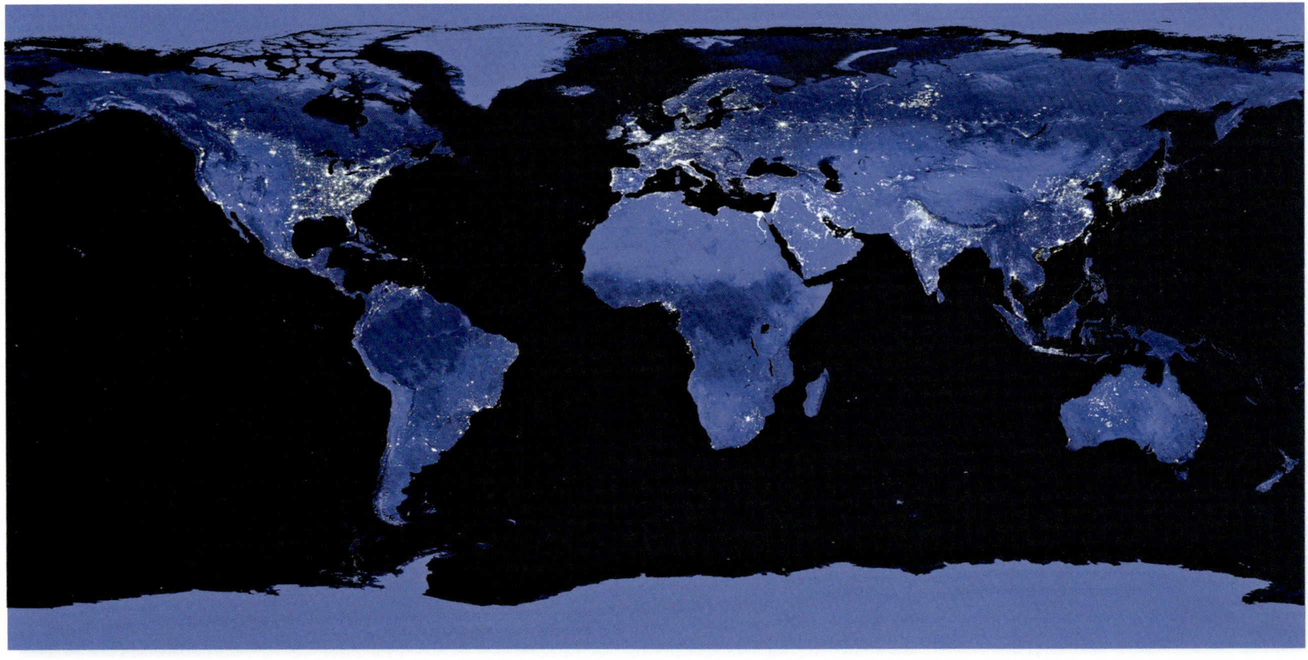

During the last half-century, the hinterland has largely disappeared from urban theoretical discourse, or has been relegated to mere background status. Under conditions of accelerated geo-economic integration, splintering national economies, the rollout of neoliberal austerity programmes, cascading social, financial and ecological crises, and proliferating local growth initiatives, cities are increasingly viewed as self-propelled economic engines. Within this post-1980s approach to the urban question, the major emphasis is on the internal preconditions, dynamics and consequences of agglomeration. Urbanisation is understood as city growth *tout court* – in effect, as *city*isation – rather than as a process that is actively supported by non-city spaces.[2]

The empty, desolate and isolated condition to which the planet's hinterlands are thereby consigned is starkly illustrated in the image of the world's night-time lights, in which brightness is treated as a proxy for cityness. This excision of the hinterland's role in urbanisation is even more starkly spatialised in the influential concept of the 'spiky world' developed by urbanist Richard Florida.[3] Here, cities are viewed as the nodal concentration points of global GDP. In both visualisations, non-city spaces appear as barren, depopulated, shapeless voids.

While the roots of this conceptualisation predate the 1970s, it was consolidated into a broadly shared episteme of urban studies following the erosion of Fordist-Keynesian, national-developmentalist capitalism. Debates on industrial clusters in the 1980s, global cities in the 1990s, postcolonial cities in the 2000s, and more recent assertions of a majority-urban world or 'urban age' represent but variations on an underlying vision of cities without hinterlands.

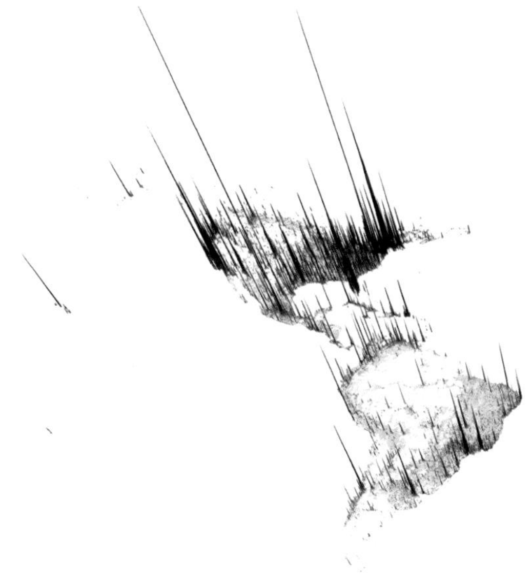

Neil Brenner and Nikos Katsikis, Visualisation representing cities and metropolitan regions as the 'spiky' concentration points for economic activities, 2010

Based on a disaggregation of national GDP data for the year 2010, this visualisation uses the approach popularised by Richard Florida in his article 'The World is Spiky' (*The Atlantic Monthly*, October 2005, pp 48–51). Hinterlands – the world's non-city spaces – are correspondingly represented as empty, barren and, by implication, economically marginal. Data source: UNEP – United Nations Environment Programme, 2012.

Counterpoint: Metabolic Urbanisation

The major contemporary counterpoints to this hegemonic, city-centric approach to urban studies are associated with various streams of urban ecological thought. Despite their otherwise divergent agendas, these dissident approaches conceive urbanisation as a sociometabolic process. From this point of view, cities are supported by diverse metabolic inputs (labour, materials, fuel, water and food) and engender a range of metabolic byproducts (waste, pollution, carbon), the vast majority of which are produced within and, eventually, absorbed back into non-city zones. Such approaches articulate a multiscalar understanding of urbanisation that encompasses not only cities and metropolitan regions, but extended landscapes of primary commodity production, logistics and waste management. Metabolic approaches to urbanisation thus seek to connect the dynamics of agglomeration to a panoply of non-city geographies – for instance, of land enclosure, population displacement, deforestation, industrial agriculture, extraction, energetics, logistics, waste processing and ecological load displacement. The most significant streams of this literature include, among others, historical investigations of city/hinterland relations, such as William Cronon's study of Chicago and the US Midwest in *Nature's Metropolis*, or Gray Brechin's investigation of urbanising California in *Imperial San Francisco*; approaches to materials flow analysis by Marina Fischer-Kowalski, Helmut Haberl and their colleagues in the Institute of Social Ecology at Klagenfurt University; the investigation of 'teleconnections' through which land-use transformations in cities impact land-use change elsewhere developed by Karen Seto and her colleagues at Yale University; and the analysis of urban ecological footprints developed by William Rees and his colleagues at the University of British Columbia.[4]

The contemporary vibrancy of metabolic approaches to urbanisation underscores the continued centrality of hinterland questions to early 21st-century urban studies. These research traditions have contributed fundamental insights that unsettle the myopic narrowing of urban investigations to cities and intercity relations, while illuminating the myriad sociomaterial processes through which city development is supported by, and actively coevolves with, non-city spaces. Thus understood, cities are *not* self-propelled. The urban process is materialised

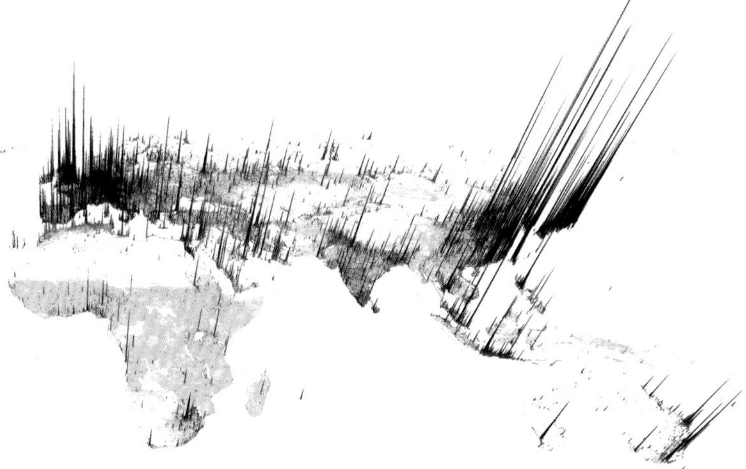

within city spaces while invariably exceeding them.[5] City and non-city landscapes are thus dialectically co-produced under modern capitalism. The urban problematique can only be deciphered adequately through an approach that systematically connects them, at once in social, political, material, infrastructural and ecological terms.

The Hinterland Enigma

Despite its role in offering powerful scholarly counterpoints to the ideology of the self-propelled city, the bulk of contemporary urban ecological scholarship has confronted the hinterland question only indirectly. While studies of urban metabolism have exhaustively quantified the material and energetic flows that mediate city/hinterland relations, they have tended to bypass the question of how non-city spaces are reconfigured through these mediations. Consequently, the hinterland itself has remained something of a 'black box': metabolic flows move in and out, but what actually happens 'inside' the box, and how the latter has itself evolved, are not interrogated. The hinterland's internal political-economic operations, land-use matrices, property relations, spatiotemporal dynamics and socioecological crisis-tendencies thus remain enigmatic.

Many contemporary urban researchers appear to presuppose a conception of the hinterland that is derived from the mercantile period of capitalist development in which von Thünen constructed his famous account of the 'isolated state' (1826).[6] Here, the hinterland is territorially contiguous with and directly linked to the city, which in turn serves as its market outlet and its manufacturing centre. Although commodity production is generalised, there is no structural impulsion to enhance labour productivity or to maximise crop yields. In this model, the non-city zone is, by definition, nonindustrial; land-use sorting occurs due to differential transport costs.

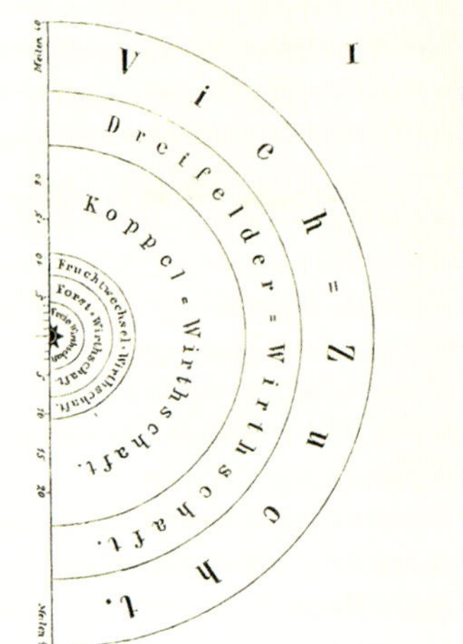

Johann Heinrich von Thünen, Model of city/hinterland relations under mercantile capitalism, 1826

Published in his *Der isolierte Staat in Beziehung auf Landwirtschaft und Nationalökonomie* (Friedrich Perthes, Hamburg, 1826), von Thünen's model shaped many subsequent generations of scholarship in urban economic geography. However, except in a few limit-cases of continued, dense metabolic interchange between settlements and their immediately contiguous supply zones, its basic assumptions have been superseded through the forward-motion of capitalist industrialisation.

The point here is not to assert that contemporary urbanists self-consciously embrace von Thünen's conception of a contiguous, nonindustrial hinterland, but to suggest that some version of this 19th-century model continues to shape our collective imagination of non-city landscapes, which are thereby reduced to an amorphous 'ghost acreage' of 'emptied spaces, homogeneous blanks yet to be inscribed by human history'.[7] As a result, scholars have only rarely sought to decipher the specific patterns and pathways through which hinterlands have been creatively destroyed since the 1850s, even though such transformations have been as far-reaching as those that are commonly ascribed to the crisis-riven remaking of cities' own built environments during the same period. Investigating such mutations will require new conceptualisations of city/hinterland matrices in relation to emergent geographies and ecologies of planetary urbanisation.[8]

Neil Brenner and Nikos Katsikis, Map visualisation juxtaposing a demarcation of the world's metropolitan agglomerations onto a rendering of the entire planet's total 'used area' at the beginning of the 21st century

Metropolitan agglomerations are shown in red and the planet's 'used area' is shown in black and grey. Agglomeration zones constitute only a miniscule percentage of the planet's operationalised landscapes, which are mostly devoted to primary commodity production (agricultural cultivation, grazing, forestry), resource extraction, logistics and waste disposal. Data sources: European Commission Joint Research Center, 2016, Global Human Settlement Layer; K-H Erb, V Gaube, F Krausmann, C Plutzar, A Bondeau and H Haberl, 'A Comprehensive Global 5 Min Resolution Land-Use Dataset for the Year 2000 Consistent with National Census Data', *Journal of Land Use Science*, 2 (3), 2007, pp 191–224; and Vector Map Level 0 (VMap0) dataset released by the National Imagery and Mapping Agency (NIMA), 1997.

Hinterlands of the Capitalocene

How, then, to conceptualise the role of hinterlands in supporting and buffering the metabolic dynamics, rifts and crisis-tendencies of urbanisation under capitalism? This challenge is, on the one hand, a conceptual one insofar as it requires us to rethink the very nature of hinterlands in the age of capital, or 'Capitalocene'.[9] It is, equally, one that will require critical appropriations of newly available sources of geospatial data, which may offer a powerful basis for investigating the contemporary rearticulation of land uses, built and unbuilt environments, and political ecologies around the world.[10]

It is not sufficient to posit that such non-city 'outsides' are constitutively important for city-building processes, or to focus on measuring the role of such spaces as 'taps' and 'sinks' for the metabolic dynamics of capitalist urbanisation. While this vast planetary hinterland covers nearly 70 per cent of the earth's terrestrial surface, and is densely layered with productive, extractive, circulatory and informational infrastructure, it has remained an obscure background to the study of contemporary urbanisation. It is precisely in this sense that the 'black box' of the hinterland must be opened and systematically rearticulated to the central agendas of urban studies. What is required is a framework that can connect historically and geographically specific forms of city and non-city space as coproduced, coevolving moments within the combined, uneven, variegated and crisis-riven world-ecologies of capitalist urbanisation.

The development of such a framework requires systematic elaboration elsewhere. Here, it must suffice to offer some initial generalisations regarding four key mutations of city/hinterland relations that have been particularly pronounced during the last half-century. These relatively abstract propositions are not intended to foreclose more contextually embedded lines of enquiry, but to stimulate further reflection, investigation and debate regarding the restlessly churning dynamics of planetary urbanisation.

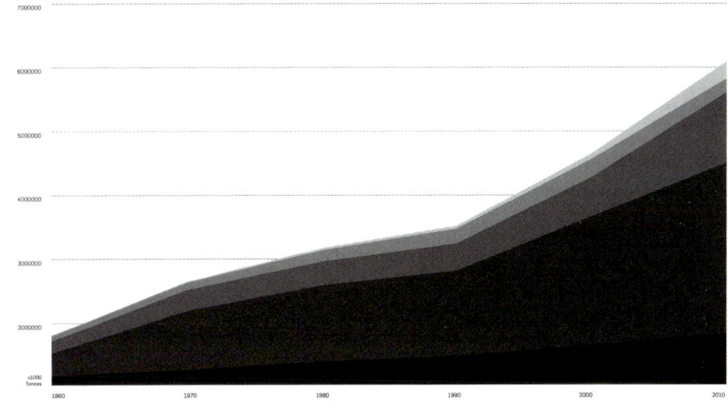

Neil Brenner and Nikos Katsikis, Visualisation of global trade of basic materials, 1960–2010

Over the last decades, the global trade in primary commodities – such as agricultural and forestry products (biomass), fossil fuels, industrial minerals, metals and construction materials – has increased more than threefold. This reflects the increasing globalisation of hinterland economies. Data source: F Krausmann, S Gingrich, N Eisenmenger, K-H Erb, H Haberl and M Fischer-Kowalski, 'Growth in Global Materials Use, GDP and Population During the 20th Century', *Ecological Economics*, 68 (10), 2009, pp 2696–705.

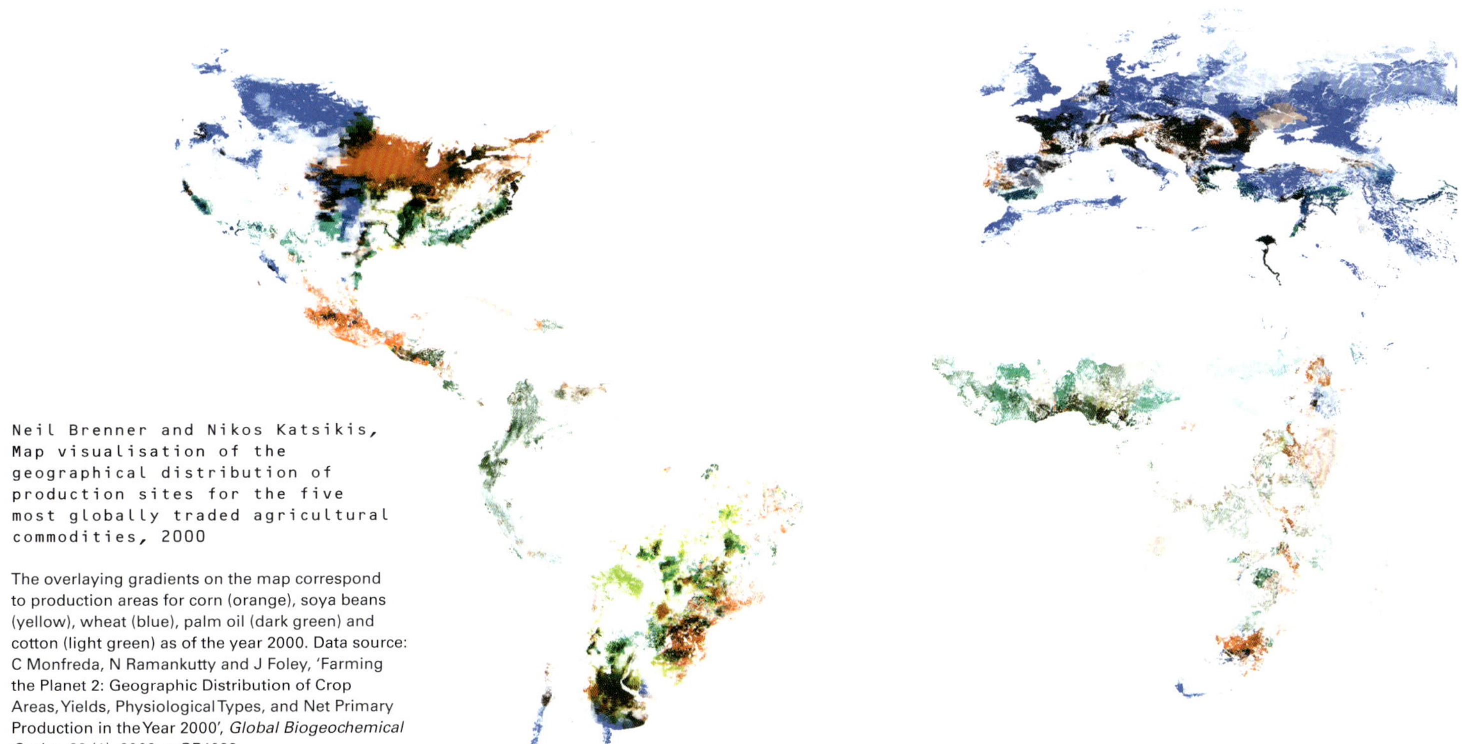

Neil Brenner and Nikos Katsikis, Map visualisation of the geographical distribution of production sites for the five most globally traded agricultural commodities, 2000

The overlaying gradients on the map correspond to production areas for corn (orange), soya beans (yellow), wheat (blue), palm oil (dark green) and cotton (light green) as of the year 2000. Data source: C Monfreda, N Ramankutty and J Foley, 'Farming the Planet 2: Geographic Distribution of Crop Areas, Yields, Physiological Types, and Net Primary Production in the Year 2000', *Global Biogeochemical Cycles*, 22 (1), 2009, p GB1022.

Distanciation and Infrastructuralisation

First, primary commodity production has been globalised and specialised, causing local, contiguous hinterlands to be enmeshed within specialised, export-oriented transnational production networks. Contiguous hinterlands remain important, but are no longer the norm, either in the older industrialised world or in most Southern megacities. This implosion-explosion of hinterland zones has been animated by capital's drive to increase labour productivity and extend interspatial connectivity, both of which entail the construction of large-scale infrastructural configurations.[11] While such strategies may temporarily boost profits, they also increase the organic composition of capital, as living labour is replaced by machinery, equipment and infrastructure. This leads to the precipitous decline of the non-city workforce ('depeasantisation'), accompanied by the social and cultural hollowing-out of rural regions, the establishment of robotised, monofunctional landscapes, and massive ecological devastation as parts of the countryside become 'sacrifice zones' for capital.

Hinterlands of Hinterlands

Second, as they are embedded within global supply chains, hinterlands lose their articulation to specific zones of direct consumption, urban or otherwise. The linear directionality of von Thünen's classic model – in which each hinterland has 'its' city, and each city 'its' own hinterland – is thus no longer a reliable guide. The point is not simply that contemporary cities' hinterlands are more distantiated than previously, but that their operational logics, infrastructural configuration, metabolic relays and developmental dynamics have been qualitatively transformed. On the one hand, most of the world's most productive, specialised and export-oriented hinterlands circulate their outputs to a multitude of metropolitan agglomerations, or across the global metropolitan network as a whole. Just as importantly, many zones of primary commodity production are now most directly articulated not to major cities and metropolitan regions, but to other productive landscapes of cultivation, extraction, processing and distribution, which are in turn embedded and intermeshed within an intercontinental logistics space. This situation is exemplified in the monocrop soya-bean landscapes of Amazonia, whose outputs are mostly exported as cattle feed to Chinese livestock hinterlands; in the export of phosphate fertiliser from Central Florida to Brazilian agro-industrial hinterlands; or in the use of hydroelectric dams to power the extractive hinterlands of northern Chile.

From Formal to Real Subsumption

Third, most forms of primary commodity production have remained heavily contingent upon the extrahuman geographies of the earth system (for instance, soil and weather conditions, water availability, or resource deposits) which can only be modified through significant industrial investment (for instance, in fertiliser, greenhouses, irrigation systems and other sociotechnical 'fixes'). Historically, therefore, the industrial operationalisation of hinterland spaces has occurred through strategies to establish new resource frontiers and, as the latter are exhausted, through compensatory efforts to intensify techno-extractive logics.

In both moments of this process, new industrial infrastructures are established and intensively operationalised before being superseded through capital's restless sociotechnical dynamism. Many contemporary hinterlands, therefore, are no longer zones of mere 'formal subsumption' in which inherited socioecological resources are appropriated as commodities for external market exchange. Insofar as the geographies and ecologies of non-city zones have themselves been systematically redesigned in order to intensify and accelerate capital's turnover time, a 'real subsumption' of hinterland spaces appears to be under way.[12] In this manner, many erstwhile hinterlands, or parts thereof, are transformed into configurations of large-scale territorial-ecological machinery: mechanised assemblages of human and nonhuman infrastructure oriented towards capital accumulation within a planet-encompassing profit-matrix.

Metabolic Rifts and Cycles of Creative Destruction

Fourth, the proliferation of specialised, capital-intensive, infrastructurally elaborate and globally interdependent zones of primary commodity production reveals not only the ways in which inherited human and nonhuman landscapes have been commodified, but the progressive exhaustion of their capacity to contribute 'ecological surpluses' to sustain and stimulate the accumulation process.[13] The proliferation of such metabolic rifts further accelerates capital's drive to mechanise hinterland geographies, at once through the substitution of manufactured inputs into the production process and through the construction of colossal techno-infrastructural configurations.[14] The hinterlands of the Capitalocene are, therefore, chronically unstable.

As ecological surpluses are exhausted, the resultant metabolic rifts severely destabilise prevalent regimes of accumulation. Consequently, established hinterland infrastructures are rendered obsolete, even though their sociotechnical capacities may have been only partially amortised. This leads to intense struggles over the choreography, form, social impacts, ecological costs and future pathways of landscape and territorial transformation.

Neil Brenner and Nikos Katsikis, Map visualisation of food, feed and biofuel cropland areas, 2000

opposite top: The overlaying gradients on this composite map correspond to cropland areas dedicated to food production (blue) and to feed or non-food uses, such as energy and industrial inputs (red) as of the year 2000. Insofar as they supply specific industrial inputs to other hinterlands (for example, cattle feed to livestock production zones, or biofuel to the energy sector), the red zones represent hinterlands of hinterlands. Data source: E Cassidy, P West, J Gerber and J Foley, 'Redefining Agricultural Yields: From Tonnes to People Nourished Per Hectare', *Environmental Research Letters*, 8 (3), 2013, p 034015.

Neil Brenner and Nikos Katsikis, Map visualisation of population densities and expansion of agricultural production zones, 1800 and 2000

opposite middle and bottom: The overlaying gradients on this composite map depict worldwide population densities (blue) and the distribution of agricultural production zones (red) between the years 1800 (top) and 2000 (bottom). Data source: HYDE 3.1 Spatially Explicit Database of Human Induced Land Use Change Over the Past 12,000 Years (2011).

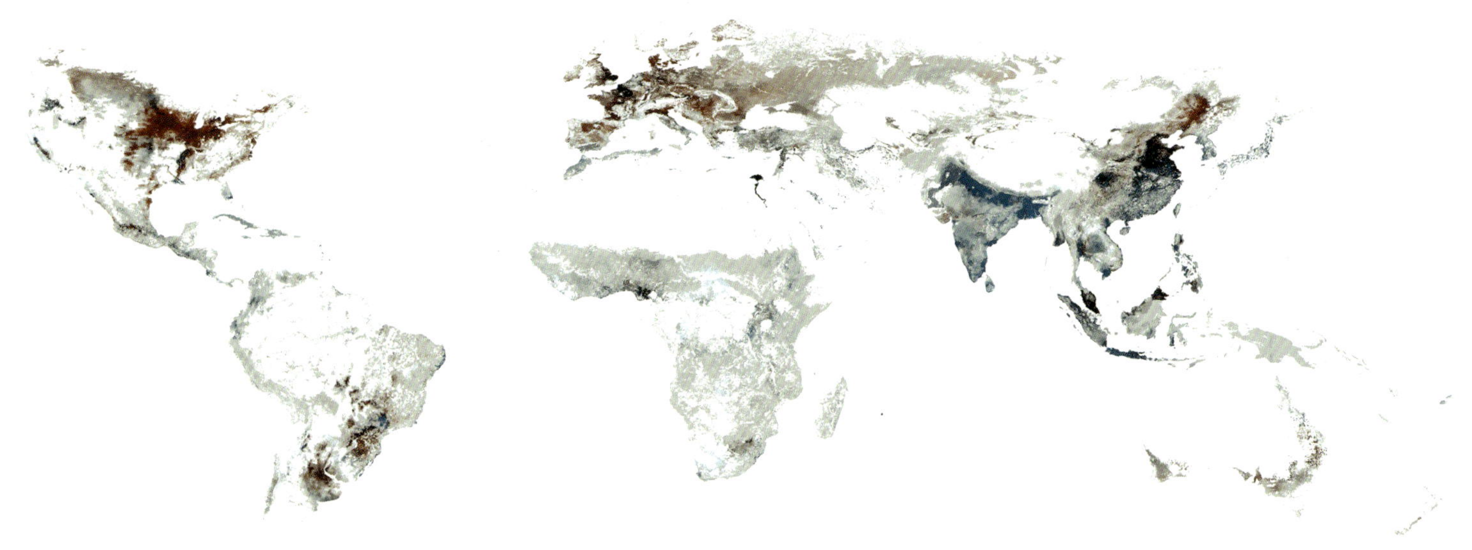

As ecological surpluses are exhausted, the resultant metabolic rifts severely destabilise prevalent regimes of accumulation

Will the violent, profit-driven illogics of planetary urbanisation continue to degrade, erode and destroy the fabric of social, political and ecological existence?

For this reason, any contemporary approach to the hinterland question must consider the systemic vulnerabilities of those non-city spaces that have been forged to support the globalising, profit-maximising dynamics of supply-chain capitalism.

The Hinterland Question, Reframed

Under contemporary conditions, there is no singular hinterland of 'the' city. Instead, non-city productive landscapes have become more specialised, infrastructurally dense and industrially intensive, and they are intermeshed with one another through extended material, operational and informational linkages, as well as through their continuous but largely indirect exchanges with (strategic nodes within) the global metropolitan network. However, the operational landscapes of planetary urbanisation are hardly a stable foundation for territorial development, social reproduction or ecological security. Indeed, even as they support enhanced industrial productivity and the

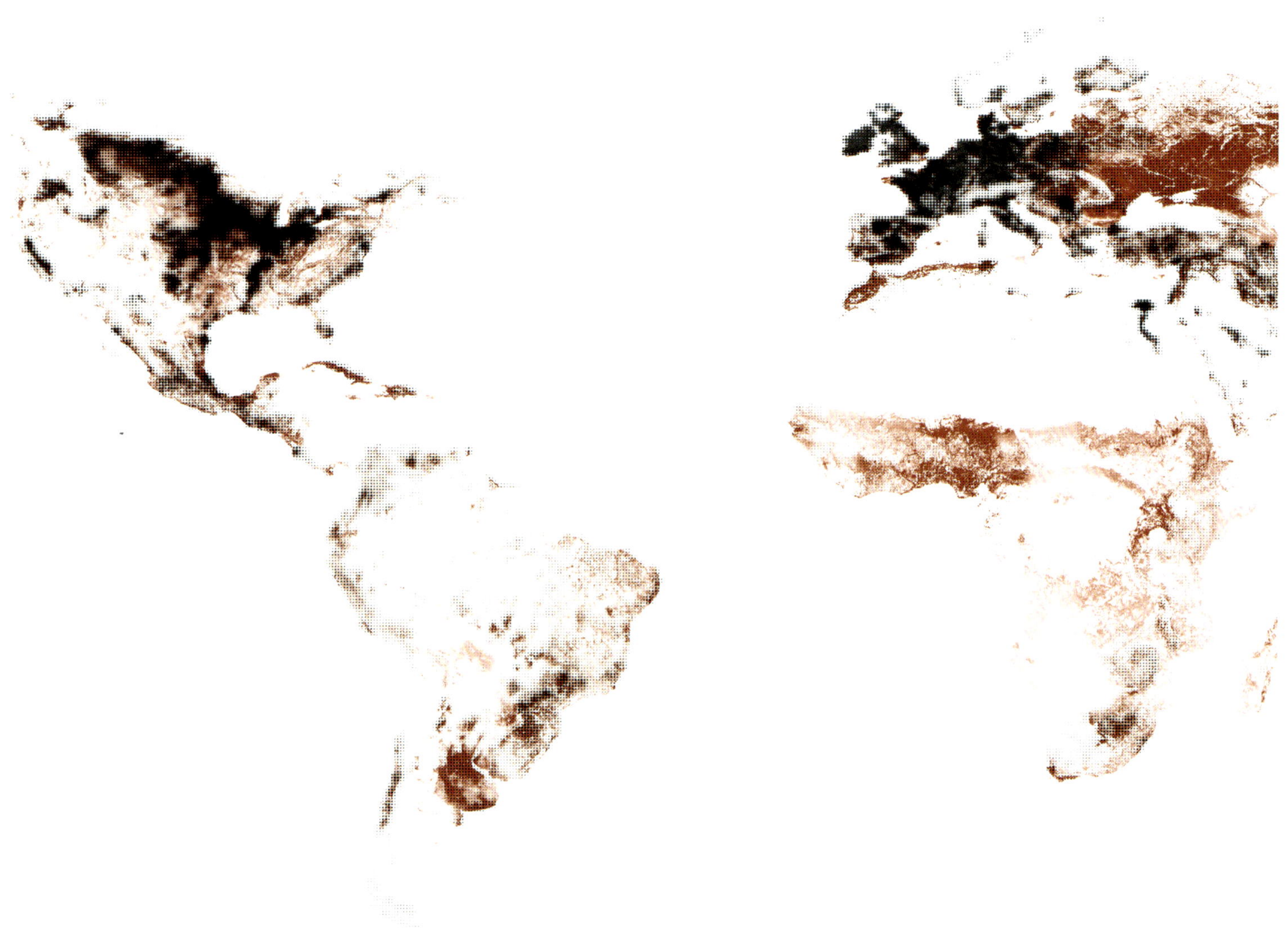

accelerated, long-distance circulation of commodities, the hinterlands of the Capitalocene expose local territories and communities to increasing turbulence, risk and precarity, while systematically degrading the ecological preconditions of both human and nonhuman life.

How, and by whom, has this planetary urban fabric been forged? What are its social, political, institutional, regulatory and ecological preconditions? What are its major contradictions, crisis-tendencies and vulnerabilities? Can the massive sociotechnical capacities it has unleashed somehow be harnessed to support more just, democratic, nonviolent, culturally vibrant and ecologically sane forms of collective existence? Are there alternative forms of urbanisation, planetary or otherwise, and can their sociometabolic dynamics be reflexively designed, negotiated and institutionalised through political agency? Or will the violent, profit-driven illogics of planetary urbanisation continue to degrade, erode and destroy the fabric of social, political and ecological existence? These are among the most urgent dimensions of the hinterland question in the Capitalocene. ᴆ

Notes
1. For an overview see: Nikos Katsikis, *From Hinterland to Hinterglobe: Urbanization as Geographical Organization*, Doctor of Design (DDes) thesis, Graduate School of Design (GSD), Harvard University (Cambridge, MA), 2016.
2. Edward W Soja, *Postmetropolis*, Blackwell (Oxford), 2000.
3. Richard Florida, 'The World is Spiky', *The Atlantic Monthly*, October 2005, pp 48–51.
4. For overviews and detailed citations of these literatures, see Katsikis, *From Hinterland to Hinterglobe, op cit*.
5. See Henri Lefebvre, *The Urban Revolution* [1970], trans Robert Bononno, University of Minnesota Press (Minneapolis), 2003; and Neil Brenner (ed), *Implosions/Explosions: Towards a Study of Planetary Urbanization*, Jovis (Berlin), 2013.
6. Johann Heinrich von Thünen, *Von Thünen's Isolated State* [1826], trans Carla M Wartenberg, ed Peter Hall, Pergamon Press (Oxford), 1966.
7. Gavin Bridge, 'Resource Triumphalism: Postindustrial Narratives of Primary Commodity Production', *Environment and Planning A*, 33, 2001, p 2154.
8. See Neil Brenner and Christian Schmid, 'Towards a New Epistemology of the Urban?', *CITY*, 19 (2–3), 2015, pp 151–82; Katsikis, *From Hinterland to Hinterglobe, op cit;* and Brenner, *Implosions/Explosions, op cit*. These texts explain in more detail the specific conceptualisation of planetary urbanisation we are presupposing here.
9. On the 'Capitalocene', see Jason W Moore (ed), *Anthropocene or Capitalocene?*, PM Press (Oakland, CA), 2018.
10. Neil Brenner and Nikos Katsikis, *Is the World Urban? Towards a Critique of Geospatial Ideology*, Actar (Barcelona), forthcoming 2020.
11. David Harvey, *The Limits to Capital* [1982], Verso (London), 2018.
12. William Boyd, W Scott Prudham and Rachel A Shurman, 'Industrial Dynamics and the Problem of Nature', *Society and Natural Resources*, 14 (7), 2001, pp 555–70.
13. Jason W Moore, *Capitalism in the Web of Life*, Verso (London), 2015.
14. David Goodman, Bernardo Sorj and John Wilkinson, *From Farming to Biotechnology: A Theory of Agro-Industrial Development*, Blackwell (New York), 1987.

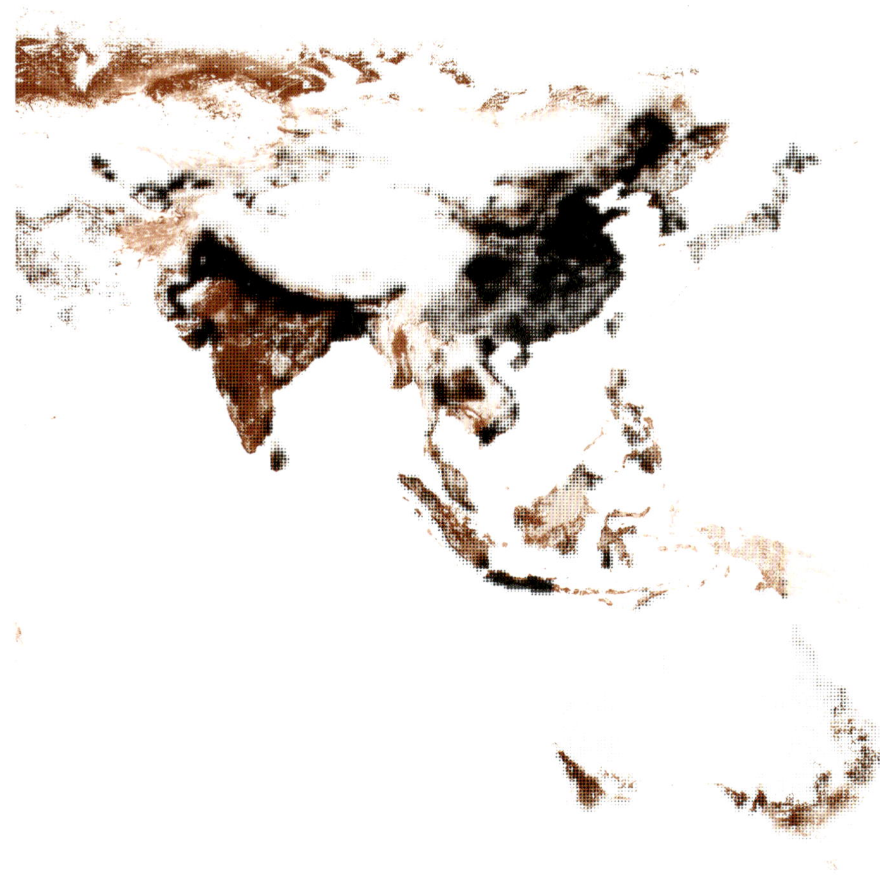

Neil Brenner and Nikos Katsikis, Map visualisation of nitrogen fertiliser use in relation to the distribution of cropland areas worldwide, 2000

Since 1950, fertiliser use has increased ninefold, while total cropland area has expanded by less than 30 per cent. This composite map depicts annual levels and locations of nitrogen fertiliser use (black-dotted pattern) in relation to the global distribution of cropland zones (red gradient). Data sources: N Ramankutty, AT Evan, C Monfreda and JA Foley, *Global Agricultural Lands: Croplands, 2000*, SEDAC (Palisades, NY), 2010, and P Potter, N Ramankutty, EM Bennett and SD Donner, *Global Fertilizer and Manure, Version 1: Nitrogen Fertilizer Application*, SEDAC (Palisades, NY), 2012.

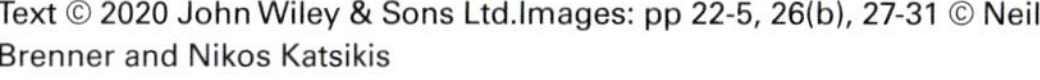

Trash Peaks

A Terrarium of the Anthropocene

Rania Ghosn and
El Hadi Jazairy

DESIGN EARTH's speculative projects make waste visible, creating public awareness of its huge ecological impact. Co-founders **Rania Ghosn**, associate professor of architecture and urbanism at the Massachusetts Institute of Technology, and **El Hadi Jazairy**, associate professor of architecture at the University of Michigan, liken their approach to shifting our attention away from just the stage to the whole theatrical machine.

In the early 1830s, frustrated by the death of his plants in polluted East London, surgeon and amateur botanist Nathaniel Ward devised a closely glazed case that set off the craze for the small glass structure – later termed 'terrarium' – as part of a Victorian popularisation of earth sciences. The aesthetic contrast offered by these lush miniature gardens to the coal-burning, soot-particle, smoke-infused environments of contemporary industrial cities served as a constant reminder of an increasingly fraught relationship to the world. If Ward's aim was to encase a small parcel of purified air, what might a terrarium look like that accepted the environmental conditions he was trying to keep at bay – the distinctive body of geological strata now accumulating, with potential to be preserved into the far future – or what is commonly referred to as the Anthropocene?

A terrarium of the Anthropocene might help conceptualise an increasingly troubled relationship with the Earth and draw attention to future fossil strata – municipal solid waste, construction debris, chemical leaks and e-waste depositories. In order to design with heaps and stacks of waste, the new landscapist must become conscious of the layers of matter entombed in the Earth's crust and the influence that they have in modifying the surface of the planet. Design must describe the world hidden within the layers of the earth in order to make visible and speculative the connections and changes across the subsoil, surface and topography of the planet. The terrarium-section becomes a heuristic device that incorporates the artificial, mutable, dissonant and parasitic as components of environments and politics.

How can design reclaim waste management systems – their forms, technologies, economies and logistics – in the politics of nature and urbanism? The practice DESIGN EARTH has called upon the geographic imagination to shift public debates on waste away from their focus on what Bruno Latour calls 'matters of fact': that is, positivist solutions to managerial crises that insist that garbage must be kept out of sight as a factually repugnant entity.[1] Latour uses the contrast between 'matters of fact' and 'matters of concern' to propose changes in the nature of evidence and its operative role. Rather than managing perpetual disappearance, 'a matter of concern', Latour explains, 'is what happens to a matter of fact when you add to it its whole scenography, much like you would do by shifting your attention from the stage to the whole machinery of a theatre'.[2] Matters of concern not only stage the issue but also destabilise how the whole construct holds together.

In the book *Geographies of Trash* (2015), DESIGN EARTH developed a four-step methodology that articulates geographic theory, spatial representation across scales, design speculation and material re-assemblies, to approach waste as matters-of-concern for design research.[3] The book counters an urban imaginary that rests on geographic abstraction, ie the 'designed' erasure of waste systems and their displacement to an externality field, to territories that are out-of-sight-beyond-accountability. By visualising the territories of municipal waste, the design research makes visible that there is no mythical outside in which the unwanted consequences of industrial and economic life could disappear from view.[4] Beyond a mere diagram of the system, the act of design speculates on worlds in which 'staying with the trouble'[5] appears as the only ethical option for technological and environmental mattering. Finally, and once the black box of systems has been exploded, a material aesthetic device draws things together into a new assemblage that re-forms both aesthetic and political consciousness.

In this line of work, the project *Trash Peaks*, which was commissioned for the inaugural Seoul Biennale of Architecture and Urbanism (2017), placed the undesired matters of waste within the geographic imagination of the city and its surrounding landscape.[6] The installation consisted of three artefacts: a printed carpet, a folding screen in six silk panels, and 3D-printed ceramic tableware. The carpet was an infographical diagram of waste streams

The E-Fungi Volcano: a multi-level sorting facility that employs fungi to extract precious metals from electronic waste.

accounting for their material, political, economic and spatial attributes, in particular as they take form in six sites of concern in Seoul. The folding screen assembled six speculative trash landform projects into one drawing that appropriated the *irworobongdo* – literally 'painting of the sun, moon and the five peaks'. Channelling the symbolic power of a motif traditionally set behind the king's throne, the recomposed landscape invited a re-valuation of the waste landscape. The three-dimensional models in ceramics – chopsticks, serving bowl, stacked plates, flask, salt- and peppershaker – were the tableware for a tea ceremony when visitors could discuss the projects. Together, these artefacts visualised, speculated on and assembled a world that re-forms and re-values waste management practices.

The Carpet: Infographics Representation

The carpet was an infographics diagram of Seoul's waste management system that listed and related the parts and processes necessary for the operation of the six identified sites. Upon close examination, the visitor discovered a vast network of entities – technologies, sites, actors and regulatory bodies – that stitched across the field to the central knot of the 37,843 tons of waste that is produced daily in Seoul.[7] The history of waste management in Seoul had been heavily reliant on sanitary landfills, which included Nanjido Island and the Sudokwon Landfill, until the city diversified its strategy to extend the lifespan of landfills and reduce waste, including recycling, isolation of waste streams, and resource recovery.

The Folding Screen: Speculative Project

A key element in domestic and royal environments of the Korean peninsula, *irworobongdo* folding screens depict an abstract symmetrical scenography of mountain peaks and the two celestial bodies of the sun and the moon. The *Trash Peaks* folding screen constructed a new symbolic landscape that assembled six stylised cyborg peaks, each a speculative project on waste technologies: landfilling, recycling, burning, re-using and reducing. The sheer artificiality of the trash peaks disclosed the tactics of the terrarium – cunning and cosmetics – to construct an amplified worldview.

The first of the six trash peaks was the Plastisphere, which recycles obsolete plastic waste and extrudes it into a plastidome that hovers on top of the Changsin hilltop toy market. Large diaphragms, called lungs, regulate sunlight and air pollution to create this tropical biome. The second was the E-Fungi Volcano: rather than sending e-waste to graveyards in Africa, this peak deploys fungi species over five levels of extraction to mine different rare earth metals from the e-waste stream in the 45-acre (18-hectare) decommissioned Yongsan Electronics Market. Third was the Janus Bukhansan incinerator, embedded into the Bukhansan Mountain, which burns waste from decommissioned landfills to power a quarter of the metropolis's energy demands. From the city, a faint trace of smoke hints at the mountain's history as a signal beacon. Looking away from the city, the inverted pyramid emerges as a monument to modern waste management.

Fourth was the Towering Construction – a spiralling tower that wraps around Mount Namsan and is made from

Tea ceremony featuring a folding screen, carpet and ceramic tableware.

The Platisphere: a hovering bio-dome environment extruded from recycled plastics.

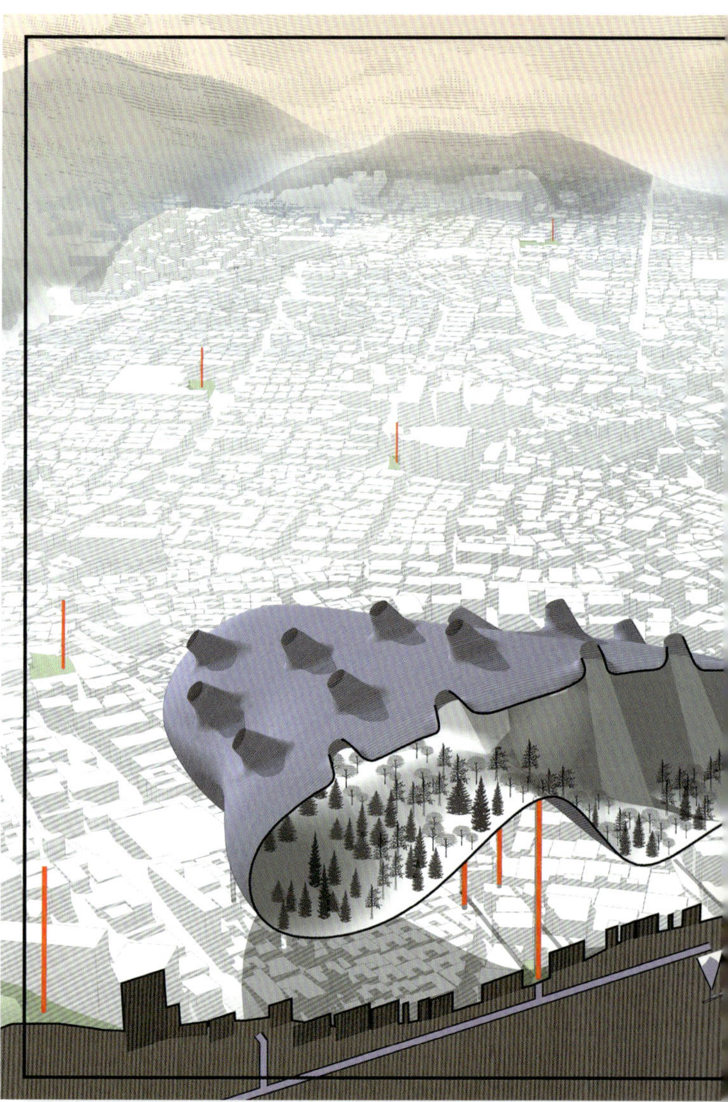

From the city, a faint trace of smoke hints at the mountain's history as a signal beacon. Looking away from the city, the inverted pyramid emerges as a monument to modern waste management

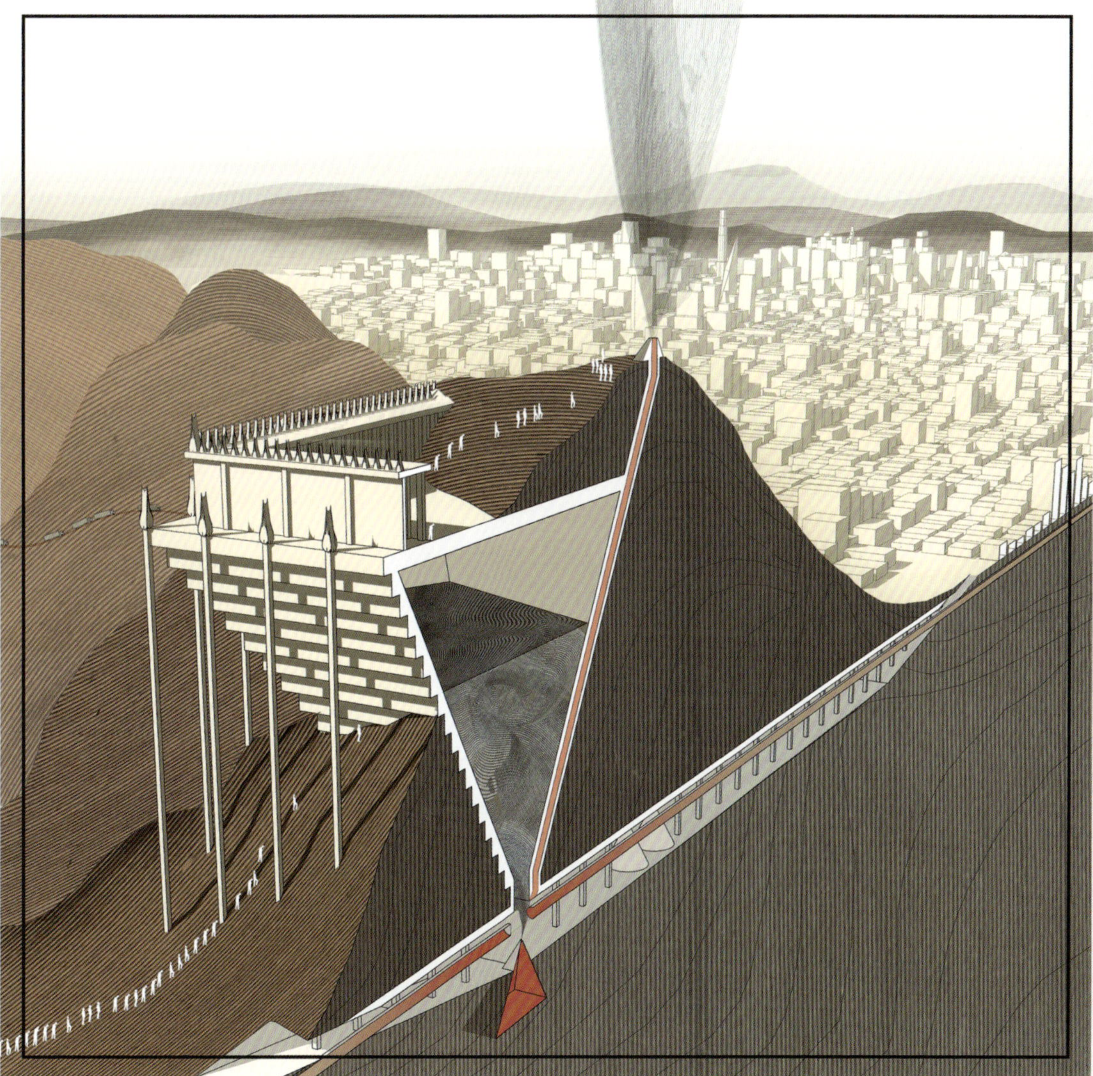

The Janus Bukhansan: a waste-to-energy facility embedded and dissimulated into one of Seoul's peaks.

The Towering Construction: a spiral tower that wraps Mount Namsan with the city's construction waste.

The Leachate Cenotaph: monumental spheres that filter organic contaminants from leaching ponds on Sudokwon Landfill.

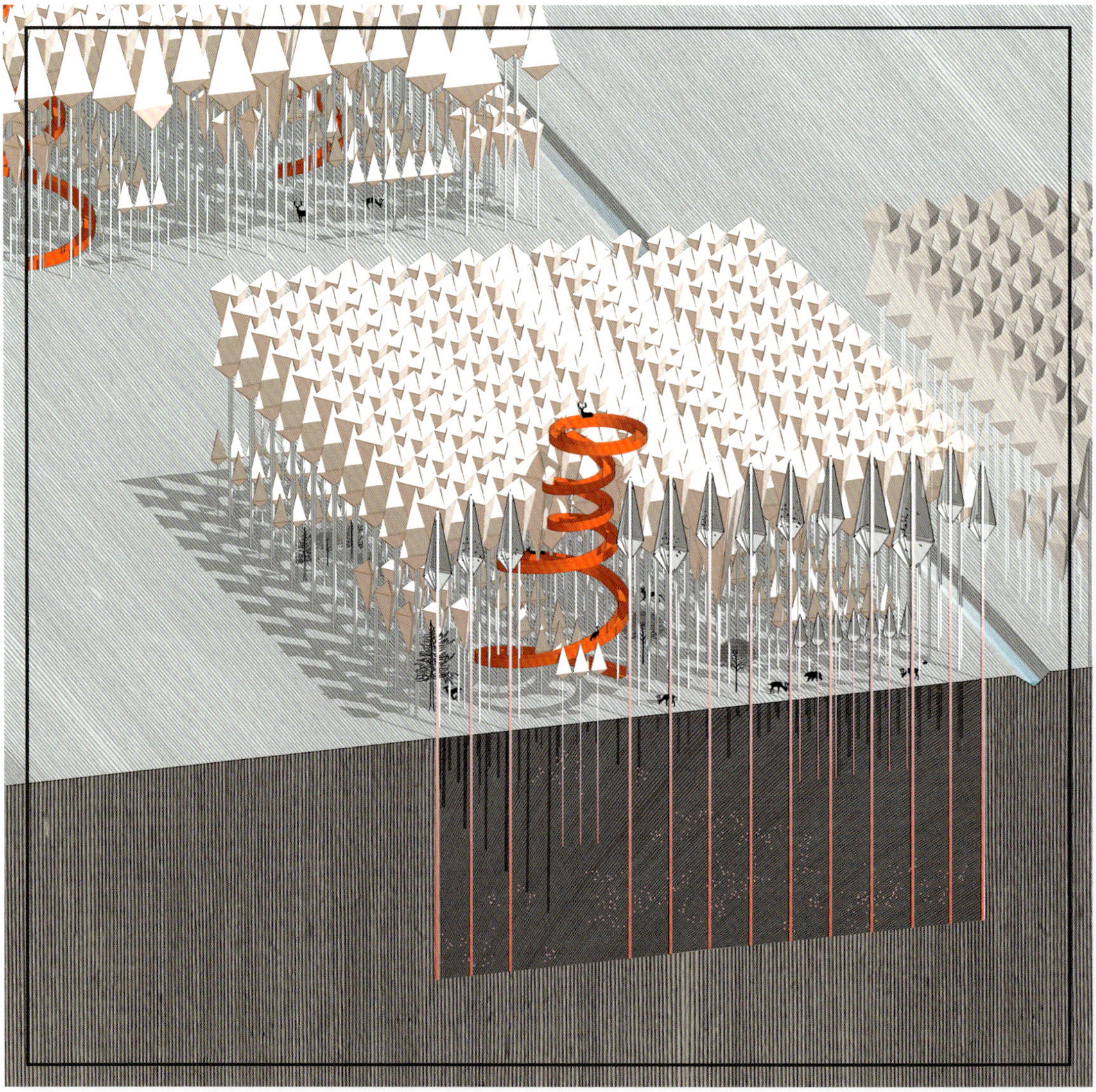

The Methane Aviary: a series of retrofitted methane pipes that become bird habitats.

All across the 20 million square metres (215 million square feet) of the reclaimed Sudokwon Landfill, a series of colossal tanks filter organic contaminant and ammonia concentrations in an environment of reflecting pools that form the fifth element: the Leachate Cenotaph

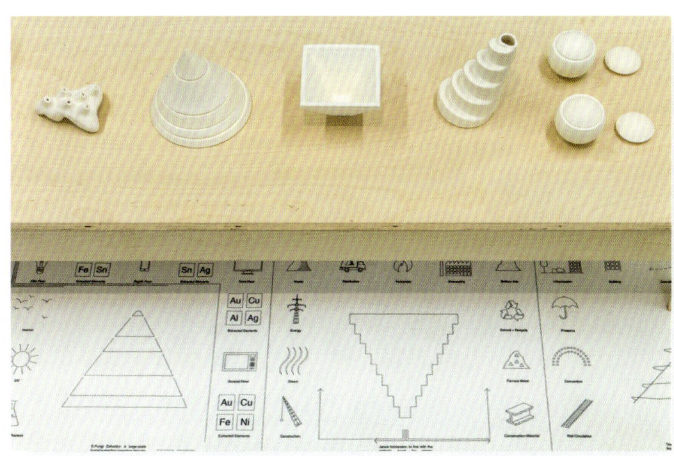

Installation detail with ceramic tableware and infographics carpet. From left to right: Platisphere salt and pepper shaker; E-Fungi Volcano nested bowls; Janus Bukhansan serving bowl; Towering Construction teapot; and Leachate Cenotaph lidded teacups.

Notes

1. Bruno Latour, 'Why Has Critique Run Out of Steam? From Matters of Fact to Matters of Concern', *Critical Inquiry*, 30 (2), 2004, pp 225–48.
2. Bruno Latour, 'Spinoza Lecture II: The Aesthetics of Matters of Concern', in *What is the Style of Matters of Concern? Two Lectures in Empirical Philosophy*, Van Gorcum (Assen), 2005, pp 27–50, at p 39: http://bruno-latour.fr/sites/default/files/97-SPINOZA-GB.pdf.
3. Rania Ghosn and El Hadi Jazairy, *Geographies of Trash*, Actar (New York), 2015. See also the projects *Belly of a Mountain* and *Neck of the Moon* in Rania Ghosn and El Hadi Jazairy, *Geostories: Another Architecture for the Environment*, Actar (New York), 2018.
4. Bruno Latour, *Politics of Nature: How to Bring the Sciences into Democracy*, Harvard University Press (Cambridge, MA), 2004, p 58.
5. Donna J Haraway, *Staying with the Trouble: Making Kin in the Chthulucene*, Duke University Press (Durham, NC), 2016.
6. *Trash Peaks* was later exhibited in 'Eco-Visionaries: Art and Architecture after the Anthropocene', curated by Mariana Pastena and Pedro Gadanho at the Museum of Art, Architecture and Technology, Lisbon, April–October 2018.
7. 'Seoul, A Resource-Recirculating City', Seoul Metropolitan Government, 20 June 2015 (updated 31 October 2016): https://seoulsolution.kr/en/content/seoul-resource-recirculating-city.
8. Peter Arnell and Ted Bickford (eds), *Aldo Rossi: Buildings and Projects*, Rizzoli (New York), 1985, p 249.
9. Gaston Bachelard, *The Poetics of Space* [1957], trans Maria Jolas, Orion Press (New York), 1964, p 152.
10. *Ibid*, p 154.

Seoul's construction waste, which constitutes 71 per cent of the daily waste generated in the city. Within the mount, three buried chambers built of concrete, steel and wood hold demolished buildings: the preserved Gyeongbokgung Palace and South Korea National Assembly. All across the 20 million square metres (215 million square feet) of the reclaimed Sudokwon Landfill, a series of colossal tanks filter organic contaminant and ammonia concentrations in an environment of reflecting pools that form the fifth element: the Leachate Cenotaph.

In 1993, the Nanjido Island Landfill was closed and its 100-metre (330-foot) high mountain of refuse excavated to ground level in order to extract valuable metal ore from within. The remaining vertical pipes, which had channelled gas from the buried waste, are retrofitted with the sixth element – a Methane Aviary for curious bird forms that have evolved to live in the hydrocarbon condition.

Tea Set Models: Public Assembled

The project models were ceramic tableware placed in a diorama relationship with the screen and that comprised the six elements: a Platisphere set of salt and pepper shakers; an E-Fungi Volcano five-piece nested bowl set; a Janus Bukhansan serving bowl; a Towering Construction teapot; a Leachate Cenotaph covered teacup with removable lid; and a Methane Aviary chopstick field. Not dissimilar to Aldo Rossi's drawings of coffeepots in the foreground of cropped city views, the ceramic models were 'miniatures of the fantastic architectures'.[8] As Gaston Bachelard reminds us, miniature objects hold the power to manifest ideas that the naked eye otherwise struggles to see or comprehend. 'Values,' he suggests, 'become engulfed in miniature, and miniature causes men to dream.'[9] Beyond a symbolic surrogate of the totality, the microcosm is also world building. It channels wonder and marvel to build a world governed by values other than those of the economy.

In the face of large-scale environmental devastation, a terrarium of the Anthropocene might dovetail with what Bachelard sees as the miniature's capacity for extracting 'large issues from small'.[10] Within a web of waste management relationships, the terrarium brought undesired matter, remote scales and systemic concepts to the personal realm. The geographic imagination tackled head-on the complex topographies of culture as they intersect with the forces of nature and technology. *Trash Peaks* came at the re-form of techno-environments through a symbolic landscape. Carefully posed, with props and scenery, *Trash Peaks* appropriated existing cultural symbols and rituals to unsettle the cognitive infrastructure of nature and technology upon which failing environmental practices rest to produce new aesthetic and political assemblies in the context of climate crisis. Simultaneously familiar and disorienting, the speculative terrarium of *Trash Peaks* stayed with the trouble to uphold that aesthetic re-form is an agency for political action. It offered unfettered entry into the landscapes of the found, fabricated and fabulated worlds in which we live, and which we have power to refashion through our creative, collective imaginings. ᗡ

INWOOD'S GEOFOLLIES

AND OTHER WITNESSES OF DISSONANCE

The Inwood neighbourhood of Manhattan is a place of nuances, of possibilities, of soft boundaries – a landscape of differing times. Landscape architect **Tiago Torres-Campos**, an associate professor at the Rhode Island School of Design, employs a variety of research-by-design methodologies to explore the potentialities of the area through the construction of a 'geologic' landscape of follies.

Tiago Torres-Campos,
Inwood's Geofollies,
Manhattan,
New York City,
2019

Inwood Hill Park's contextual conditions.
Common perceptions of the park as a natural haven and
a relic of another time are simplistic and inaccurate.

Inwood's Geofollies is a critical design exploration that emerges from within a wider research-by-design project. The broader research was initiated in 2014 and its main intention is to examine how Manhattan's geologic conditions may help inform alternative representations of the island-territory.

The actions involved in thinking geologically about landscape and architectural practices include a 'thinking with' as well as a 'feeling through' the idea of 'geologic'.[1] In opposition to 'geological', defined as an adjective which qualifies something relating to or based on geology, 'geologic' may be defined as a set of enmeshed relational conditions that emanate their own aesthetic and cultural sensations.

The conceptual proposal focuses on Inwood Hill Park to test ways in which landscape architectural practices may witness, register and evoke material dissonance.

A Park Amongst Visions of Eco-Inaccuracy

Inwood Hill Park stands on the northern tip of Manhattan, spreading downhill from the top of a big rock that is in direct contact with the junction of the Hudson and the Harlem rivers. The hilly landscape reveals steep slopes facing north and gentler slopes facing the city to the south. It is covered in a dense deciduous forest linked with a tidal salt marsh through a thick and dark valley. The rocky hill was forged by the glacial retreat of the Wisconsin Ice Sheet that once covered a big part of the North American plate.[2]

Offering views across the Hudson to the Palisades in New Jersey and to other hills in the vicinity, the park was once an important settlement of the indigenous Lenni Lenape people. It was also the place where Peter Minuit, Director General of the Dutch North-American colony of New Netherland, allegedly purchased the island in 1626 from the local tribes. These communities called the island *Mannahatta*, often translated as 'the land of many hills', 'place for gathering wood to make bows', 'the island where we all became intoxicated' or, simply, 'island'.[3] When speculating on the insular territory before the establishment of the first colonies, landscape ecologist Eric W Sanderson refers to a luxuriant and very diverse landscape, crossed by many streams that nourished fertile valleys, wetlands and forests.[4]

There is a common perception amongst contemporary Manhattanites that Inwood is the last piece of native forest and salt marsh on the island, a relic from another time. For today's local inhabitants, the urban-eco amateurs, the occasional tourists, and the groups of pupils and university students who visit it on a daily basis, the park is described as a natural haven. The bowl-shaped valley conceals the busy city and opens up to a glimpse of a tidal riverscape. This ecological vision of the park is, however, too simplistic and inaccurate.

Inwood's Dissonant Occupancies

If the Anthropocene can be conceived also as a 'politically infused geology',[5] a contemporary situation where materials are vibrant witnesses and active generators of processes and aesthetic sensations,[6] then Inwood's landscape material inscriptions may testify dissonant occupancies and register geological violence enacted by political, sociocultural and extractive actions. Disparate histories and stories of rock, salt and woodland gain preponderance throughout centuries of a city growing northwards through displacement, dispossession, demolition and cleansing. Rock, salt and woodland inform a material palette inscribed in and erased from the landscape multiple times.

From a deforested landscape during the 18th-century wars between the British Empire and the rebellious colonies, to a place of many asylums and weekend retreats for the rich merchants coming from New York's downtown in the 19th and early 20th centuries, Inwood Hill has experienced significant landscape change.[7] Following the archaeological works which uncovered several caves that had once served as dwellings for the Lenni Lenape, the city first considered the possibility of creating a park here in the 1930s. The influential city planner Robert Moses used his position as Park Commissioner to initiate a process of demolishing the mansions and asylums that occupied the ridges.[8] The old, fractured geology bears witness to the many uses and occupancies related with sheltering and leisure, as well as forced treatment or incarceration.

With the surrounding rivers revealing meaningful tidal variations at this point around the hill, salt gradients used to define most of the boundaries between land and water. Inwood, with the last salt marsh in Manhattan, evokes what once were valuable tidal habitats feeding the island, such as reed, kelp or oysters. But, in fact, the existing retaining wall that protects the park from the marshland's cyclic flooding equally registers the process of city growth which progressively replaced all soft waterscapes on the island with clear-cut port-like edges.

Most woodlands in Manhattan were harvested, either to give place to urbanisation in times of peace, or for timber production for guns and fortifications in times of war. Some were also replanted. Moses's decision to erase the cultural landscape and replant the woodland that currently exists in the park is perhaps one of the most recent actions in a long succession of plantation and harvesting on this hill. While it decisively contributed to the park's value as an ecological haven, these actions may also be described as an act of White cleansing based on an abusive power relation in the city against human as well as nonhuman vulnerable minorities who once inhabited the hill.

Moses's resolution in the 1930s was closely aligned with the new planning policy to create public outdoor spaces throughout the city. As the grid carved its way across Manhattan's topography, several parks of different sizes were defined in areas where the rock was either too big or too hard to cut. A closer critical cartographic analysis of the several plans for the northern boroughs developed along the early decades of the 20th century reveals that many options of bringing the city grid up Inwood Hill were considered.

The creation of a park in Inwood reflects some of the wider geopolitical dynamics behind the creation of Manhattan's park systems, especially north of Central Park, where geology is harder, and therefore more visible and determining than in the island's lower parts. The majority of these landscapes are limited by significant streets or avenues, which do not conform to the orthogonal grid but distort it instead. Some of these significant boundaries lie directly on top of geological faults spreading

Inwood Hill
Fort Tryon
Fort Washington
Riverside Park
Dyckman Street Fault
River Park
Jackie Robinson
Highbridge Van
St. Nicholas
Marcus Garvey
Morningside Heights
122nd Street
102nd Street
59th Street
125th Street Fault
135th Street Fault
Central Park
Park Avenue
9th Avenue

across the island and they also divide the city into its different administrative areas.[9] Inwood's geology, for example, is part of a complex system defined by the Dyckman fault, which runs beneath the street carrying the same name. The fault separates Inwood Hill and Isham Park, to the north, from the two long systems running south, one in the west composed by the parks of Fort Tryon and Fort Washington, and the other in the east composed by the parks of Harlem River, Highbridge and Jackie Robinson. Even in their apparent geological stability, when considering these rocky organisations across the vastness of deep time, it is perhaps interesting to imagine them as geological constellations moving very slowly and according to choreographies constrained by the faults' geometric rules of compression, tension, slippage or partition.

Witnesses of Dissonance

Inwood's dissonant accounts across space and time demonstrate that ecology can erase narratives of violence through enforcement, theft or war. Yet, at the same time, this landscape's current state of maturity equally nurtures meaningful collective values of preservation and conservation.

Inwood's Geofollies is a conceptual proposal which critically explores the tensions arising from these rather conflicting positions. In so doing, it proposes the redesign of the existing salt marsh as well as a series of follies that accumulate around the marshland's tidal interfaces. In this design exploration, rock, salt and woodland are once again considered as a valuable material palette, this time in contemporary configurations.

The redesign of the salt marsh focused on the removal of the existing retaining wall to unleash the tides. The design process evolved with the creation of a drawing machine, where a mechanical device was attached to a physical model of the landscape. The machine supported the exploration of some of Inwood's highly site-specific dynamics: atmosphere and tides exchanging salt; drainage of run-off and underground water along the steep geology; and finally, compression, permeability and erosion caused by gravity. The machine equally brought into the project its own operating materials, by moving atomised and liquefied paint mixed with salt across the model to test principles of fluidity usually related with landscapes of salt – saturation, crystallisation, erosion and sedimentation. As the salty ink slowly dried up by gravity and evaporation, it created a new undulating topography. These tests were then adapted to the park's topographical and hydrological conditions – in both plan and section – as means to explore new interfaces between the sloped woodland and the tidal marsh.

In parallel, there was a careful cartographic and in-situ mapping of the park's marks and scars left by the previous dissonant occupancies, from building footprints to cave dwellings, and from glens to glacial potholes. Both the architectural and the geological excavations narrate stories of geological violence mostly by their evocative material absence. In the project, these voids were conceived as their reverse – extruded geological masses. The consideration of these conceptual volumes scattered along the hillsides suggested a potential gravitational dragging downhill, similar in many ways to the erratic boulders being dragged by glacial retreat and scattered across Manhattan, some of which can still be observed in their curated positions across Central Park. The final accumulation and recombination of the conceptual masses at the foot of the hill generated new architectural assemblages, conceived as contemporary landscape follies.

Conceptual recalibration of Inwood's landscape. The park's current conditions are recalibrated according to conflicting positions between previous dissonant occupancies and the contemporary ecological value.

> # Both the architectural and the geological excavations narrate stories of geological violence mostly by their evocative material absence

opposite: Design exploration of a reinvented landscape. Rock, salt and woodland inform the redesign of the salt marsh and the proposal of follies along the tidal interfaces.

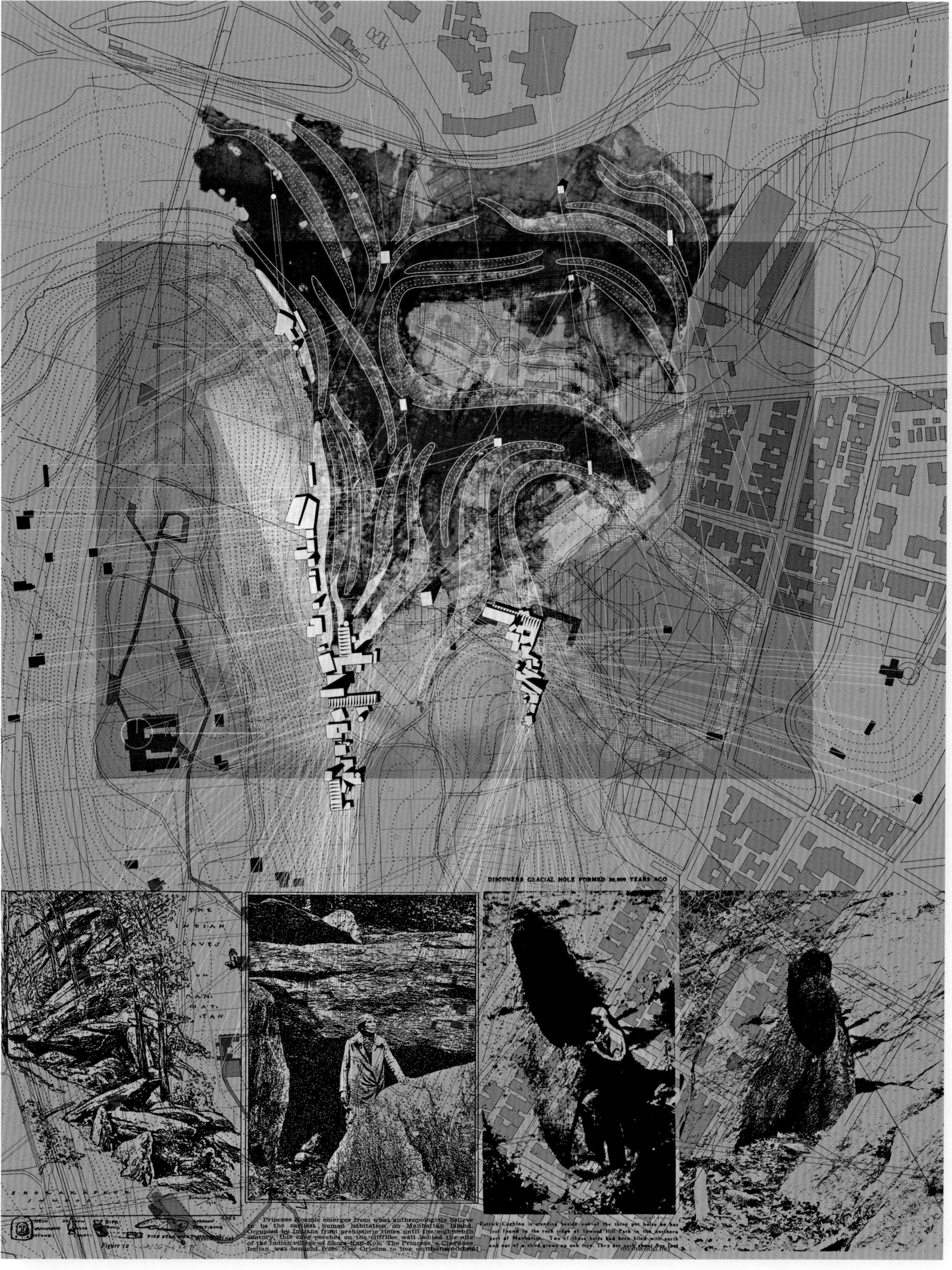

DISCOVERS GLACIAL HOLE FORMED 30,000 YEARS AGO
Princess Noamie emerges from what anthropologists believe to be the earliest human habitation on Manhattan Island. Occupied by Indians from prehistoric times until the eighteenth century, this cave perches on the clifflike wall behind the site of the Indian village of Shora-Kap-Kok. The Princess, a Cherokee Indian, was brought from New Orleans to live on
Patrick Coghlan is standing beside one of the three pot holes he has just found in the rock ridge of Inwood Hill Park in the northern part of Manhattan. Two of these holes had been filled with earth and out of a third grows an oak tree. They are
Figure 12

Salt marshland unleashed: a new landscape of salt.
The design tests principles of saturation, crystallisation,
erosion and sedimentation.

The idea of folly is meaningful here. Follies are forms distinguishable from the landscape context in which they exist and without explicit programmatic ambitions, sometimes without a programme at all. In his reflection about the Osaka Follies, the Japanese architect Arata Isozaki writes that a folly is 'a building that traditionally has no function'.[10] He adds:

> follies are small and fraught with difficulties. Designing a building normally entails responding to a specific use, or communicating a particular message. But with the folly there is no direct meaning to put across, no concrete function to fulfil. It isn't quite architecture. It isn't sculpture.[11]

The proposed Inwood follies attempt to forge relations with the existing context, geologically and topographically, but also culturally and materially. They compress and distort space and time insofar as to offer alternative narratives which can influence potential new ways of reading, understanding and experiencing landscape. The English architect Cedric Price discusses the folly's potential in promoting such a distortion:

> The true folly reaps the harvest of a silent eye, suspending truth itself in a delightful way. … The folly distorts time, place and space and in so doing mixes magic with mystery, fun with fantasy, 'now' with 'then'. Humour is essential in its creator, and wit a bonus.[12]

Formulations of a new undulating topography in section.
The proposal explores new interfaces between the sloped
woodland and the tidal marsh.

Design proposal for Inwood's Geofollies. The new architectural
assemblages result from the accumulation and recombination
of extruded volumes at the foot of the hill.

Any sense of humour in landscapes like Inwood is not easy to find, accept or explore. It is eventually coated with dissonance and the much-needed observation that some of these beautifully immersive landscapes carry with them violence of some sort. In that sense, the humorous practice is much closer to a sort of dark irony which nevertheless accepts 'serious playfulness' as a mode of thinking manifested through controversy or cynicism.[13] The Inwood follies question geological excavation and extraction, yet they are also born out of disruptive cut and fill. They question geological stability, yet they also require their own architectural solidity amongst the fluid tides. They problematise the manifestation of power in the smoothing erasure of dissonance, yet they further accentuate a series of architectural and landscape gestures driven by a personal design exploration. And, in the process of forging intimacy with the geologic, they become 'geofollies'.

The final accumulation and recombination of the conceptual masses at the foot of the hill generated new architectural assemblages, conceived as contemporary landscape follies

Forging intimacy with the geological. The proposed geofollies accept dissonance, contradiction and controversy as integral characteristics of the landscape.

Together with the unleashed marsh, which invades them with the tides and cyclically sprays them with salt, the geofollies do not assume any educational or entertaining competencies. If anything, they quietly – but not silently – problematise some of the contemporary dynamics involved in landscape conservation as actions gesturing towards alienation through forcing the ever-changing landscape to be crystallised in tamed and known forms of acceptable outdoor museology.[14]

Thinking About Landscape Geologically

In an age of attentive contextual reconnaissance, landscape architectural practices may benefit from intensifying their traditional focus on dense analytical and propositional explorations of non-binary site-specificity and nonlinear locality. These are design practices of noticing, making sense and bearing witness that slow down and attune to things that may not exist in a completely solid way, but are going through a meaningful transitional process of becoming. These practices can encourage a mode of 'thinking geologically': that is, among other things, a way of thinking about landscape as a set of inter- and intra-connected conditions that reveal – even if in a weird or flickering way – the possibility or ability for new things to happen. ᴆ

Notes
1. Heather Davis and Etienne Turpin, 'Art & Death: Lives Between the Fifth Assessment & the Sixth Extinction', in Heather Davis and Etienne Turpin (eds), *Art in the Anthropocene: Encounters Among Aesthetics, Politics, Environments and Epistemologies*, Open Humanities Press (London), 2015, p 3.
2. Charles Merguerian and J Mickey Merguerian, *Field Trip Guidebook: Isham and Inwood Parks, NYC* (Lamont-Doherty Earth Observatory, Manhattan Prong Workshop), Hofstra University Geology Department and Duke Geological Laboratory (Durham, NC), 2014.
3. Robert Juet, *Journal of Hudson's 1609 Voyage*, entry 2 October, New Netherland Museum / Half Moon (Albany, NY), 2008: http://newnetherlandmuseum.org/Juets-modified.pdf.
4. Eric W Sanderson, *Mannahatta: A Natural History of New York City*, Abrams (New York), 2009.
5. Kathryn Yusoff, *A Billion Black Anthropocenes or None*, University of Minnesota Press (Minneapolis, MN), 2019, Preface.
6. For a thorough explanation on 'vibrant matter' and the agency of object, see Jane Bennett, *Vibrant Matter: A Political Ecology of Things*, Duke University Press (Durham, NC), 2010. The concept of 'material witness' was first developed by artist and writer Susan Schuppli: https://susanschuppli.com/research/materialwitness/.
7. Judith M Fitzgerald and Robert E Loeb, 'Historical Ecology of Inwood Hill Park, Manhattan, New York', *Journal of the Torrey Botanical Society*, 135 (2), 2008, pp 281–93: http://www.jstor.org/stable/40207578.
8. Arthur H Graves, 'Inwood Park, Manhattan', *Torreya*, 30 (5), 1930, pp 117–29: http://www.jstor.org/stable/40596696.
9. Charles Merguerian, and John E Sanders, *Geology of Manhattan and the Bronx: Guidebook for On-The-Rocks*, New York Academy of Sciences (New York), 1991.
10. Arata Isozaki, 'Osaka's Green Crossroads', in Pamela Johnston and Dennis Crompton (eds), *Osaka Follies*, Architectural Association (London) / Workshop for Architecture and Urbanism (Tokyo), 1991, p 5.
11. *Ibid*.
12. Cedric Price, 'The Folly', in Johnston and Crompton, *Osaka Follies*, op cit, p 7.
13. For a thorough account of dark humour in relationship to ecological thought, see Timothy Morton, *Dark Ecology: For a Logic of Future Coexistence*, Columbia University Press (New York), 2018. For a more complete explanation of the importance of serious playfulness in challenging its direct, corporativist opposite, playful seriousness, see the podcast conversation between Timothy Morton and Sean Lally, *Night White Skies Episode 002*: https://soundcloud.com/user-561947272/ep-002_timothy-morton.
14. For a thought-provoking exploration of the museum as a space of historical alienation, see Peter Sloterdijk, 'Museum – School of Alienation' [2007], in *Art in Translation, Volume 6, Issue 4*, Bloomsbury (London), 2014, pp 437–48.

Luis Callejas and Charlotte Hansson

Pelag

Islands as a Model of the Ocean

Alph

LCLA office,
Pelagic Alphabet,
Oslo Architecture Triennale,
2016

A drawing of Märket island displaying the intricate border that was defined after the separation of Finland and Russia. The new, more egalitarian distribution of land allocated to Sweden a surface area equivalent to that occupied by the Finnish lighthouse. The story behind this conflict inspired the idea of attempting to imagine each island being split equally.

LCLA office's *Pelagic Alphabet* – an island microcosm of strange forms and imagined geographies – seeks to use the unique status and special place of the 'island' in human culture. Described here by the architecture and landscape studio's partners **Luis Callejas and Charlotte Hansson**, it is a place of blended imagination, topological exaggeration, boundary and territorial dispute.

By envisioning the bottom of the ocean as a continuation of the ridges, peaks and plateaus that form the idea of littoral landscapes, rather than an isotropic surface, it is not hard to imagine that this unfathomable space can be aestheticised as a landscape too. Islands that punch out of the surface of the ocean can be read as small models of the totality of an underwater landscape, the solid key to the interpretation of underwater features. This is not a far-fetched idea, as it is known for example that experienced sailors know how to read the landforms above water in order to imagine what the underwater terrain looks like, and more importantly, if it presents dangers for navigation in tight archipelagos.

For designers of landscapes and buildings, the relative smallness of the figure of the island also allows for it to be drawn and read as a model that embeds information about a larger unfathomable territory. The *Pelagic Alphabet* project developed by LCLA office for the 2016 Oslo Architecture Triennale began with a preparatory exercise that involved drawing a plan and elevation of an island. When these drawings were then translated into three-dimensional ceramic models, stories were inscribed in them, about conflicts that define the specific territories. Representing the island as a ceramic object charged the model of something so small with events that operate at vast scales, making them manifest and intensified in the almost architectural figure defined by the island's shore. In this case, imagining a fictional or exaggerated topography was not so much about the continuation of the ground below the water, but rather about charging topographic information with subjective details and stories about landscape conflicts such as territorial disputes between island countries.

As the project acquired fictional topographic information in elevation, while maintaining the plan as accurately as possible, non-existent topographic features and textures communicated territorial conflicts by means of exaggeration.

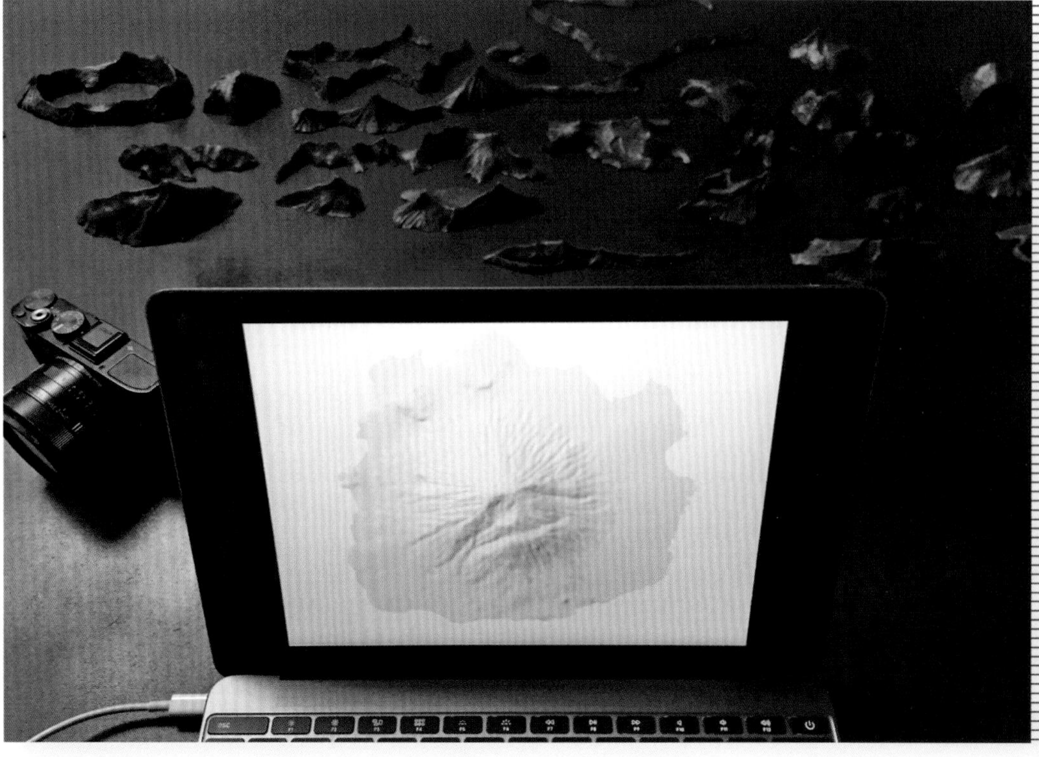

The process of translating 3D models into ceramic models and back. The precise three-dimensional models of 40 contested islands were interpreted and re-modelled in clay. Afterwards the clay models were photographed and compared with the original digital models. Topographic information was added at each stage.

Drawing an Island: A Solid Model of the Ocean

For obvious reasons, the challenges presented by the inhospitable space of the ocean do not allow us to look at the pelagic space in the same terms and with the same aesthetics we would use for land. When observing land masses, we often resort to the old tropes of scale and distance, and almost always to some form of visual distinction between foreground, middle ground and background. Distance and scale also seem to work differently in the ocean: for example, a human body has to be in top shape to be able to swim a distance of 100 metres (330 feet), while walking that same distance in a park can be performed by almost anyone. Visiting a small island is a similarly polarised experience, often involving the perils of navigation, yet once there, there is often no chance of a long walk. It is the long journey and then the contact with the sand or rocks that sticks in the memory: once again, extreme remoteness and then extreme proximity.

This contrast between extreme proximity and extreme remoteness brings specific representational challenges, as there is no middle ground to be described in the same terms we traditionally use to describe terrestrial landscapes: there is only the big picture and the small, creating an argument for representing an island as a small ceramic object.

When looking at canonical drawings of mountains, it is almost always as if they are seen from the vantage point of a fictional flat-plain middle ground, ironically almost as if viewed from the ocean: far enough to objectify the dominant shape, close enough to distinguish topography or vegetation. When examining drawings that exist between art and geographic representation, like the Prussian polymath Alexander von Humboldt's early 19th-century representations of the Andes, it is evident that the artist/naturalist chooses one aspect that needs exaggeration in order to communicate the mountain as a totality, intensifying detail for living matter and topography.

Topography as History

As designers, not naturalists, it is possible to play liberally with the representation of topography to relate it to subjective stories, rather than always attempting a precise survey. In other words, the island needs exaggeration to behave as an effective model, most islands being flat sandbanks.

The focus with the *Pelagic Alphabet* installation was therefore to play with the way we perceive islands, and to explicitly exaggerate the topography of the ones that are split between two nations. Often boundaries between countries are traced over the perceived limits defined by geographic entities such as a river or a mountain ridge. Fictitious topographic features allow a re-drawing of the borders to be imagined, exposing sovereignty as a material condition driven by the objectification of the ground.

When confronted with questions of landscape and territoriality in the ocean, and trying to engage with coastal landscapes, architects often end up resurrecting and instrumentalising the old tropes of military outposts, lighthouses and other such totems. *Pelagic Alphabet* went against the romantic image of the architectural object on an island by framing the island as an autonomous object in its own right, where the ground's texture, rather than buildings, tells a clear story about conflicts in the vast ocean.

Two stories were fundamental to the project's development. One is that of the 20th-century oceanographic cartographer Marie Tharp. If there is one person who could be called the Humboldt of the ocean, it is certainly her. The beauty of her work was that, while famous and heroic explorers were producing vertical columns of data that were hard for the public to decipher and were surprisingly missing the big picture, she was assembling the relief of the underwater world by turning those columns into a recognisable topography, a landscape, described by drawings that look like familiar representations of

The ceramic models were worked by hand, introducing variations to the topography, with a focus on texture. The models were fired in an oxide reduction environment to avoid the texture being concealed by glaze.

The new models of the 40 islands were grouped to compare the different
fictional topographies. The models were accurate in plan projection;
however, the elevations were exaggerated to describe the subjective
stories behind the territorial claims to which the islands are subject.

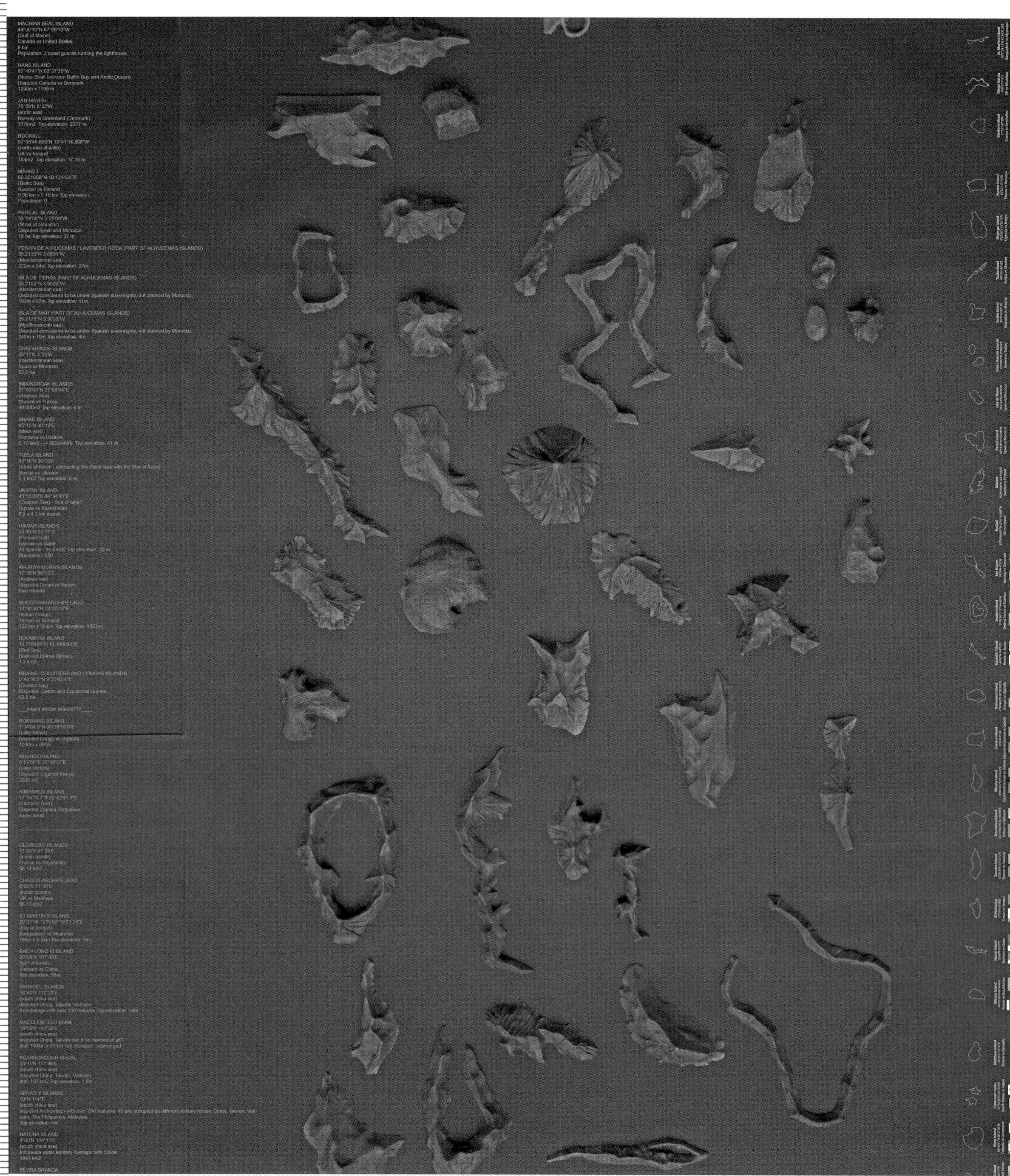

A drawing attempting to expose a possible division of land based on
the Märket island case (Finland–Sweden). Each island is represented
here with its original topography. The contours in the drawing
provided the basis for further topographic deformations.

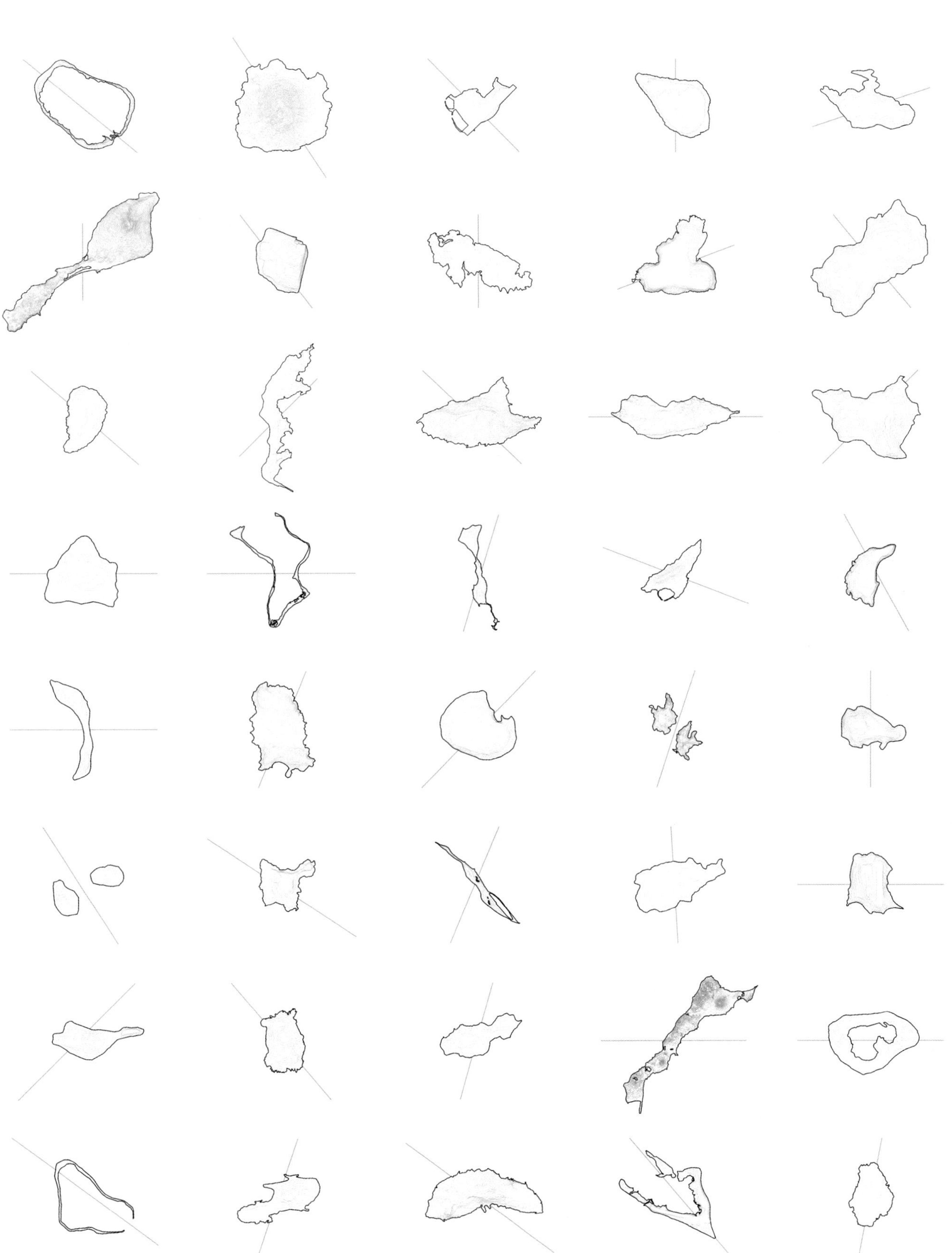

A rendered view of the reconstructed islands in 3D before translation into ceramic models. Some of the islands look rather flat at this stage, particularly the lower sand banks. The 3D digital models served as a basis for deciding where to introduce more texture or increase the angles of the slopes. Further modifications were guided either by a free interpretation of each island's story, or by the need to exaggerate texture for the model to be buildable in clay by hand.

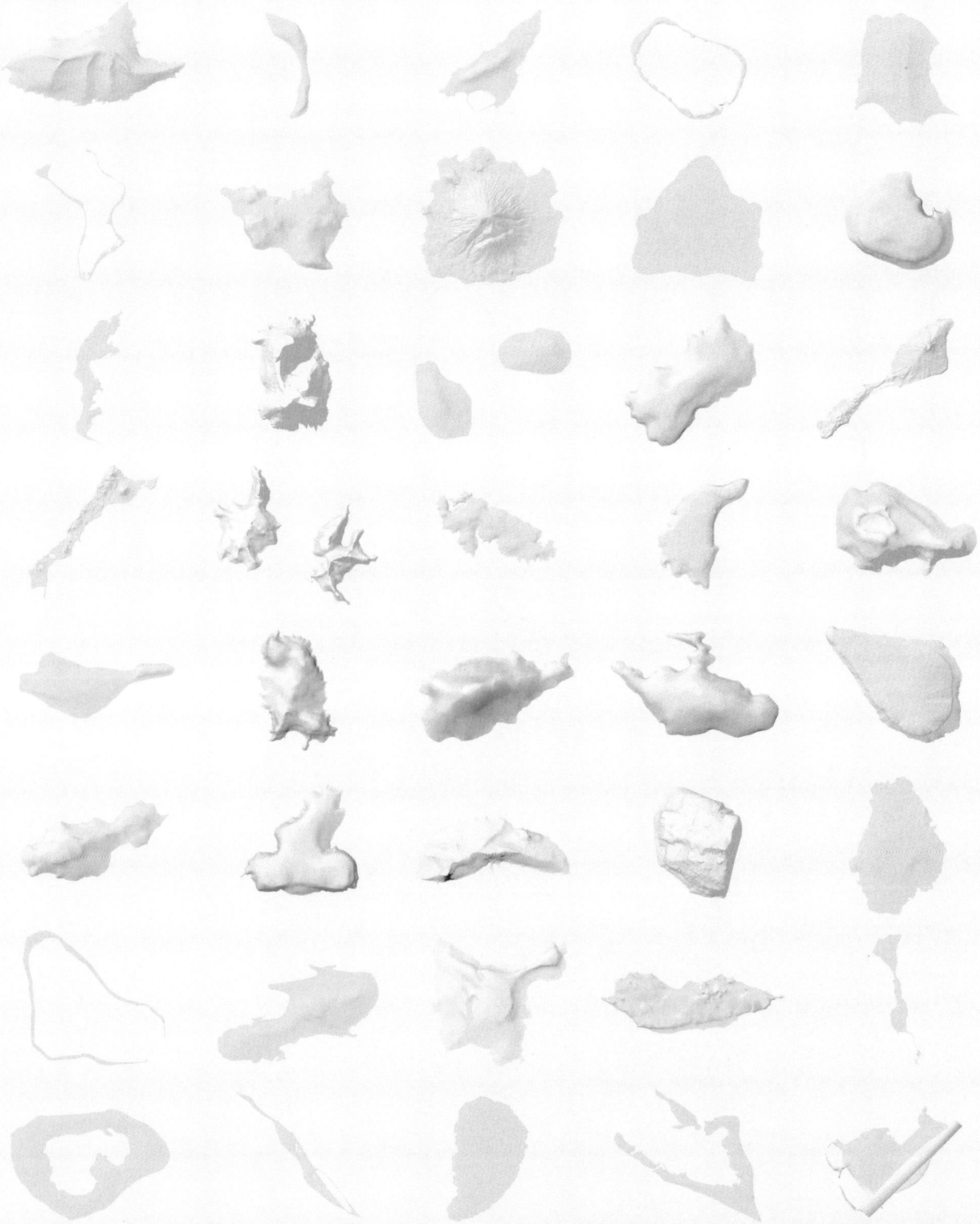

mountains, plains and valleys. In 1977 she produced the first world ocean floor panorama and by connecting the dots discovered an underwater mountain chain that is a couple of thousand kilometres (over a thousand miles) longer than the Andes. The translation of columns of information into an image allowed us to see the world afresh and derive new theories, most notably helping to reconcile opinions on the theory of continental drift.

The second story was the study of the border between Finland and Russia, and later Finland and Sweden, on Märket island. Perhaps one of the most intricate borders in the world, it was resolved by achieving an egalitarian distribution of land between Sweden and Finland by correcting the border (originally placed by the Russians) in a way that keeps the lighthouse on the Finnish side. This story raised the question of what topographic deformations would be needed in order to re-plot a border on an island with no totems or outposts that can be traced to a nation.

Littoral Modelling

During an initial research stage for *Pelagic Alphabet*, LCLA office set out to identify the ideal islands to be studied: that is, ones that are the subject of disputes between two nations. The original idea was to consider only islands with a sharp visible topography; however, it became evident that it is the flatter, more threatened sandbanks that are more vulnerable to changes in sovereignty. The research led to the pinpointing of 40 islands from around the world. The source of the conflicts is diverse, ranging from fishing rights to oil exploration to

military installations. Among the chosen examples were Clipperton (France–Mexico), El Tigre (Honduras– El Salvador) and Navassa (Haiti–US).

Each of the islands was then studied as an autonomous object. In some cases it was possible to borrow from existing maps, but for most of them there was no widely available topographic information. The team therefore manually traced contours, guessing the shapes from blurry aerial images found in different sources. This was the first act of construction of the fiction – the invention of a terrain by adding details that most likely do not exist.

The work next shifted from 2D to 3D. The digital files looked like boring, almost flat cakes, and were not produced to a consistent scale. At this point all the team wanted to see was the island as an object in relation to the story behind it. It was clear that the stories about conflicts were as biased and imprecise as the office's interpretations of topography.

The final step was about 'inventing' the islands in clay. The fundamental driving question was straightforward: What should this topography look like if this is the particular story? Each island acquired false information while retaining the original plan figure. The coast, the contact with water, was the only link to its real-life counterpart.

The conclusion drawn from the way in which this project developed is that, similar to how sailors have to imagine the bottom of the ocean by reading masses that exist outside the water, it is possible to imagine the stories behind a charged and vast aquatic landscape by studying how islands are represented and objectified. ⚏

Re-tracing the topography by studying the ceramic models. The islands were then translated back into 3D models in order to generate hyper-realistic images of the new islands. The resulting models allowed the generation of images that correspond to precise plans and fictional elevations.

Advanced Landscapes

Guest-Editor **Ed Wall** charts the history and successes of the Advanced Landscape and Urbanism group at the University of Greenwich where he is Academic Portfolio Lead, and illustrates the exploratory nature of its pedagogy.

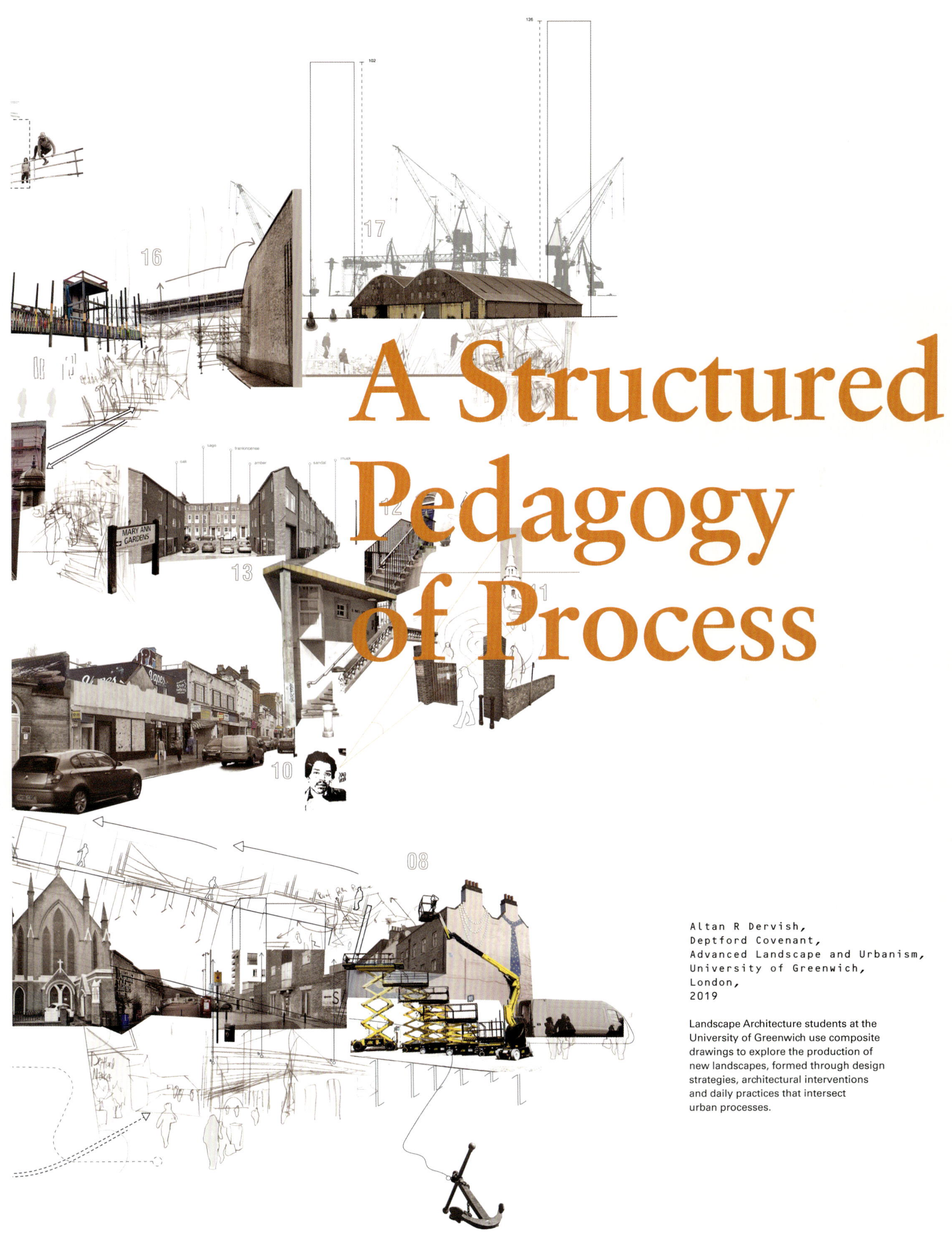

A Structured Pedagogy of Process

Altan R Dervish,
Deptford Covenant,
Advanced Landscape and Urbanism,
University of Greenwich,
London,
2019

Landscape Architecture students at the University of Greenwich use composite drawings to explore the production of new landscapes, formed through design strategies, architectural interventions and daily practices that intersect urban processes.

'As a design discipline embedded in site specific contextual analysis, contemporary landscape architecture seems to lean towards a more structured pedagogy of process.'
—Rosie Martin, MA in Landscape Architecture, University of Greenwich, 2018[1]

Advanced Landscape and Urbanism was formed in 2015 as a loose grouping of graduate students, design tutors and academics interested in the design of landscapes and cities. The aim was to activate and frame an expanding field of landscape – from within traditions of teaching landscape at the University of Greenwich, since 1965 – defined by the projects produced rather than adopting positions, restricting practices or narrowing definitions. Advanced Landscape and Urbanism began with conversations about teaching landscape, with James Fox, a design tutor in the School, asking: 'What *is* landscape at [the University of] Greenwich?' It has since become a collective platform to explore possibilities of landscape teaching and landscape research. Formed at the intersection of Master's, PhD and academic practices, it encourages new approaches to research and design experimentation that investigate the environments, tools and lives of landscapes.

The operations of this MA Landscape Architecture project proposal aim to question the practices and boundaries of urban development to establish moments of negotiation within a new landscape of socio-spatial relations.

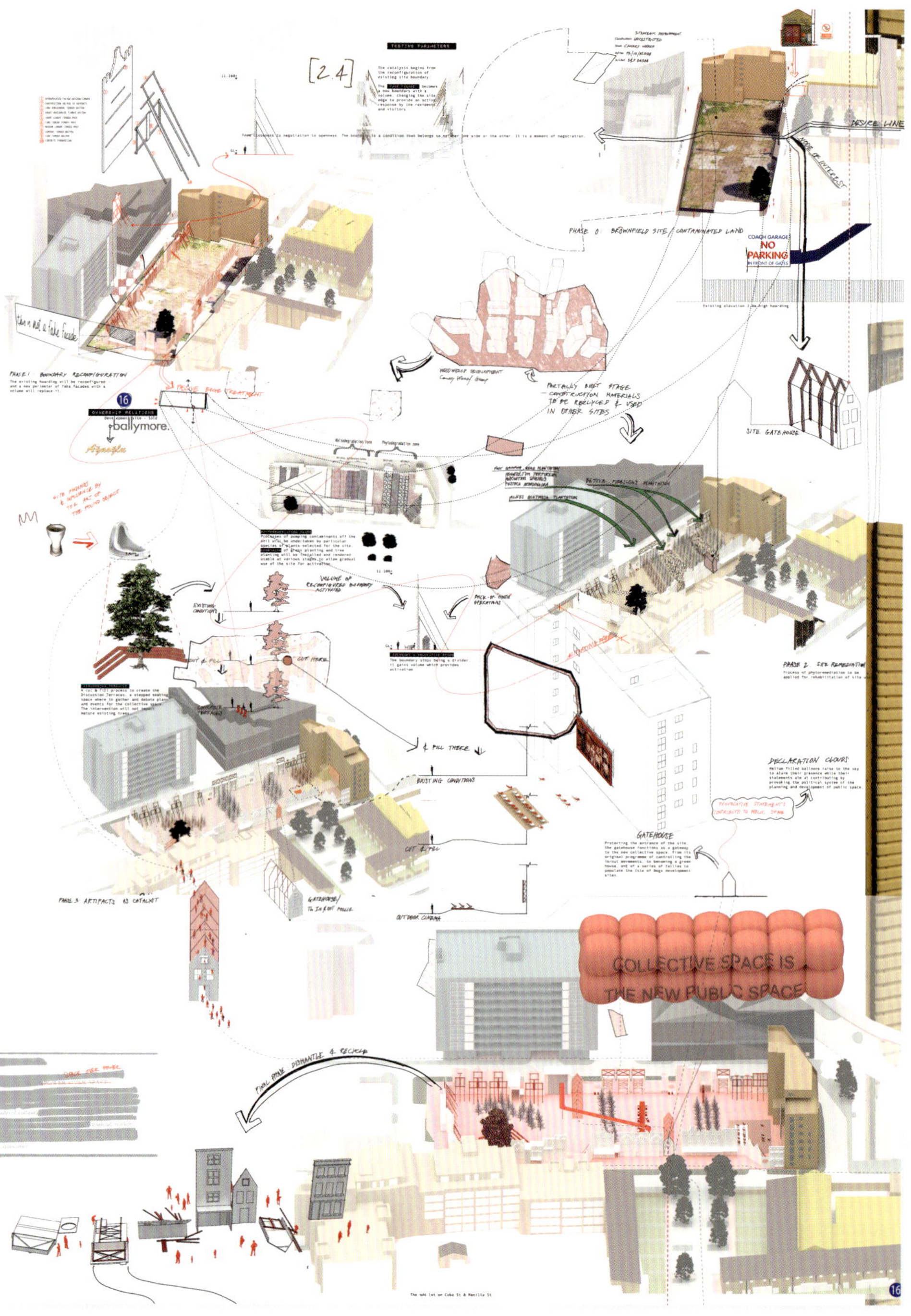

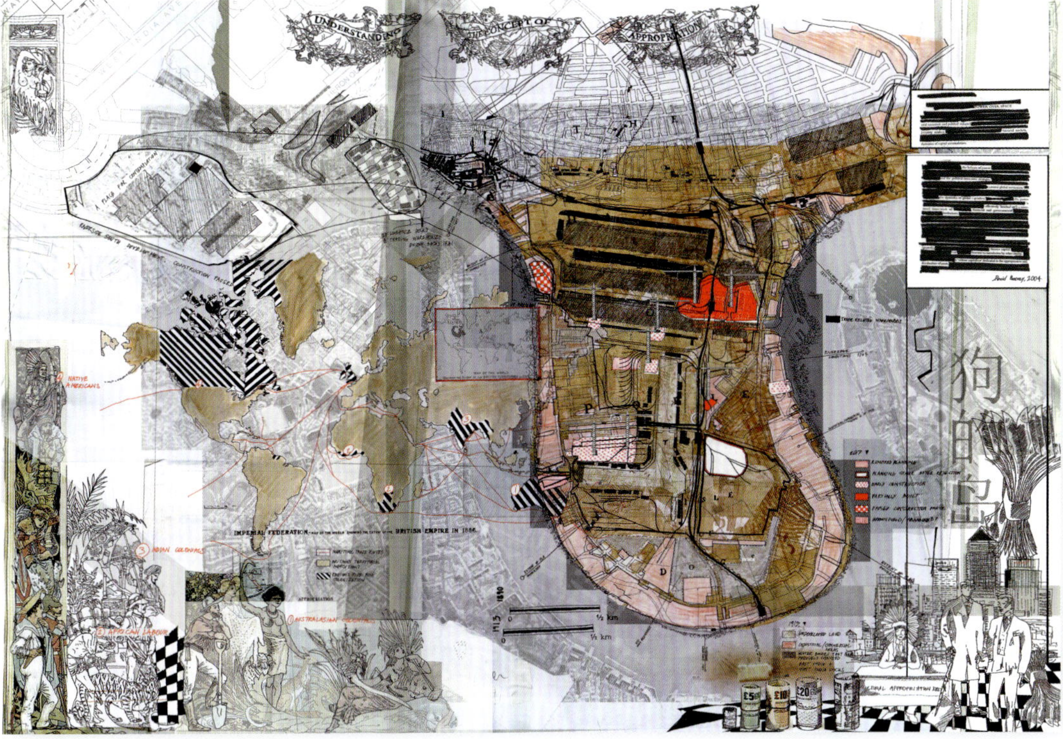

Composite research drawings developed by MA Landscape Architecture students at the University of Greenwich provide useful tools to analyse the complexities of urban landscape relations, such as in London's Isle of Dogs, where histories of colonial trade, appropriation of land and capital accumulation entangle.

Five years on, this article reflects on the investigations that have driven research projects, student designs and collective endeavours in Advanced Landscape and Urbanism. It is informed from many conversations, including a four-hour semi-structured group discussion exploring future pedagogies focused on the design and research of landscapes and cities. Advanced Landscape and Urbanism is framed in three ways. Firstly, an emphasis on landscapes being socially constructed through interrelations between individuals, people and the worlds they produce. Secondly, an embracing of diversity of landscape approaches, resulting in projects created from unique combinations of experiences, methods, skills and ambitions – but differentiated through focused research projects and structured design studios. Third, Advanced Landscape and Urbanism embraces tensions within and between landscapes, working with the complexities and contradictions that landscapes entail, as much as with the potential agency of landscape relations.

'Landscape' informs all the ways of thinking and working in Advanced Landscape and Urbanism. The term 'landscape' is used to denote entanglements of relations – landscape as physical and immaterial relations between people and the environments around them, from the immediacy of locales in London (from where most projects originate) to more distant places and less tangible times. While appreciating the urge to focus on 'what' landscape is – particularly conventions of design practice that consider landscapes as physical places to visit, study and transform, or as representations of places within visual arts and literature – projects address more critical concerns for 'how' landscapes are produced and 'who' has the opportunity to be part of these processes. Not denying the impulse to explore landscapes spatially and visually, projects give less emphasis to 'what' the products of landscape are. Designs do not ignore the significance of the physical conditions of existing landscapes to be rigorously studied or an obsessional eye to how design narratives are conceived, developed and refined. They instead emphasise the importance of critically 'working' images and texts in order to understand and represent the complexities of landscapes. Such relational approaches to landscapes, which are investigated in Contested Boundaries and the Appropriation of Space (2018) by MA Landscape Architecture student Cesare Cardia, are core to the exploration of sites and the necessary refinement of methods and designs – explored through extraordinary drawings that aspire to make exquisite marks on the ground.

Advanced Landscape and Urbanism employs the term 'Advanced' to emphasise that experimentation should be at the forefront of an expanding and creative field of landscape. From this restless position there is a constant searching for future forms, practices and questions of landscapes. Free from the necessity to realise designs within the professional field of landscape architecture and urban design – where the stakes are high in regards to what communities, clients and ecosystems can gain or lose – Advanced Landscape and Urbanism explores critical questions that investigate the conditions of landscapes, from the intimately small to the unfathomably large, addressing concerns of climate change, migration, digital technologies, spatial justice and urban growth. Helena Rivera, a design tutor in the studio, argues: 'These are projects that can only happen in the university!'[2] Working from this position also provides a critical perspective towards conventions of landscape, whether inside or outside the university, in order to further landscape knowledge and practices. The expectations of research culture for producing innovative bodies of knowledge should be no different than the demands on design practice for addressing unique conditions of landscapes in ways that may not have been undertaken before. Student projects such as Mais Kalthoum's Island Factories (2018) on the Isle of Dogs in London illustrate the potential of working with future scenarios of climate change and the responsibilities that Global North countries have to communities under threat in the Global South. Her iterative process of gathering information, synthesising and speculating – focused through extended periods of reworking – resulted in a proposal that made explicit 'how' and by 'whom' an elaborate ecology of landscapes was produced.

Mais Kalthoum,
The Island Factories,
Advanced Landscape and Urbanism,
University of Greenwich,
London,
2018

Three Island Factories are proposed by University of Greenwich MA Landscape Architecture student Mais Kalthoum for London's Isle of Dogs, creating inhabitable islands for Global South communities impacted by rising sea levels.

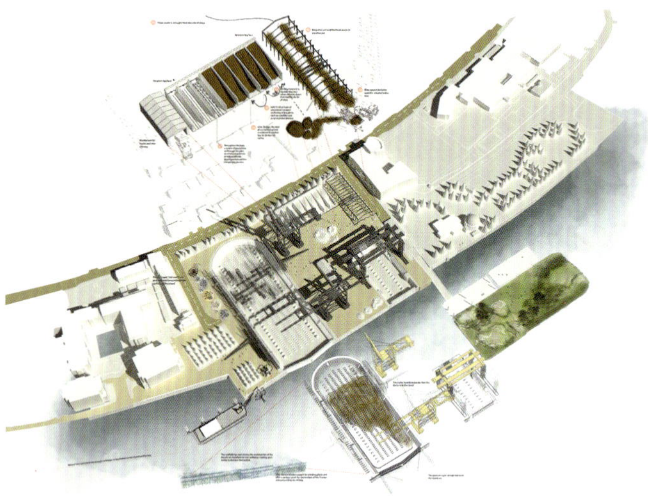

The Leisure Island Factory creates artificial floating landscapes of leisure – reminiscent of Robert Smithson's Floating Island that was drawn by the artist in 1970 but only realised by a team led by Balmori Associates in 2005 – constructed as barges and aggregated to form new inhabitable places.

New industries are proposed in the Island Factories to confront and mitigate the impacts of rising sea levels that are caused by global climate change and that make some towns, cities and countries uninhabitable.

Relations-Between

To focus on the relational dynamics of landscapes is to emphasise the significance of change and time. But while the notion of change is commonly accepted, including growth, inhabitation, erosion and decline, within the context of practice there are difficulties designing and realising dynamic landscapes. Doreen Massey, in her 2006 essay 'Landscape as Provocation: Reflections on Moving Mountains', gives emphasis to 'constant movement, the inevitability and inexorability of process (rather than entity); on flow rather than territory'.[3] While she accepts that this can be understood as 'a *conceptual* issue' and that 'in the practical conduct of the world we *do* encounter "entities"', she also argues that 'change cannot be rejected'. She continues: 'The stake is not change itself (the denial of it in the past or the refusal of it in the future), for change of some sort is inevitable; rather it is the character and the terms of that change. It is here that the politics needs to be engaged.'

Advanced Landscape and Urbanism projects question relations of ownership, occupation and development of landscapes through writing, drawings and models, attempting to visualise activities, discourses and processes that are often obscured from view or understood in more social or ecological terms. The influence of theoretical concepts, through the works of landscape architects, social scientists or political ecologists, provides essential references connected to immediacies of uneven urban development or loss of biodiversity, through empirical research and site-specific design. A collaboration that began in 2015 with Sayes Court, a Deptford community action group led by Roo Angell and Bob Bagley, has opened up conversations around practices of globalisation, arrivals of refugees and community initiatives on the River Thames. The collaboration provided the basis, in 2016, for the first issue of *Testing-Ground*,[4] a journal developed by Advanced Landscape and Urbanism, and the context for student projects such as Altan R Dervish's Deptford Covenant (2019) that confronts the intricacies of streets, journeys, buildings and development.

By working with the social constructedness of landscapes, student projects have been able to address environmental concerns (such as climate change and ecological justice) and economic issues (such as urban development and gentrification) but with a focus on what is at stake for different individuals and places entwined in these processes of change. The writings of the landscape architect Jane Hutton and of the sociologist Caroline Knowles are a significant reference for this relational understanding. In her 2013 exploration of 'reciprocal landscapes',[5] Hutton highlights the material relations between landscapes created and destroyed through design projects; while Knowles's *Flip-Flop: A Journey Through Globalisation's Backroads* (2014)[6] traces the entanglement of places and lives in the material relations of global supply and manufacturing chains. Rather than focusing on the artefacts of finished projects and the conditions of physical places, as most landscape and city design ventures do, Advanced Landscape and Urbanism gives a greater interest to the relational aspects of landscapes, such as the processes from which they are produced. While studios may be initiated from London, design projects simultaneously investigate wider geographical relations between locations of resource extraction, global flows of materials, competition between cities for development and the implications for the people and places involved. Hutton argues that 'Conceptualizing the sites of material production as integral rather than external to design would shift theoretical concerns of the landscape project without necessarily shifting its site boundary'.[7]

This approach is taken by John Joseph Watters in his Master's project for Angerstien's Wharf (2015), where the flows of materials become the basis from which a new network of public spaces and actions are proposed. Such projects, Hutton continues, have 'the potential to both examine the ways in which nonadjacent spaces are designed contiguously, but also to speculate about how these reciprocal relationships might be designed themselves'.[8] Design and research projects aim to make more visible the individuals involved, the tools that they employ (in the design process and within sites) and the varying and multiple relations that they have with their environments. Patrick Geddes's *Valley Section* drawings that emphasise relations between practices of work and landscapes of a typical valley, first published in 1909, are key references – but with greater concerns for people and places entangled in contemporary processes of urbanisation, future devices of landscape and yet-to-be-explored relations between them.

John Joseph Watters,
The Curated Celebration of
Angerstien's Aggregate Wharf,
Advanced Landscape and Urbanism,
University of Greenwich,
London,
2015

Angerstien's Wharf is Europe's largest sea-dredged aggregates terminal where huge piles of aggregate are constantly evolving and shifting amid a complex system of conveyors and structures. This MA Landscape Architecture proposal inventively brings public access and interaction in direct juxtaposition with the wharf's industrial operations.

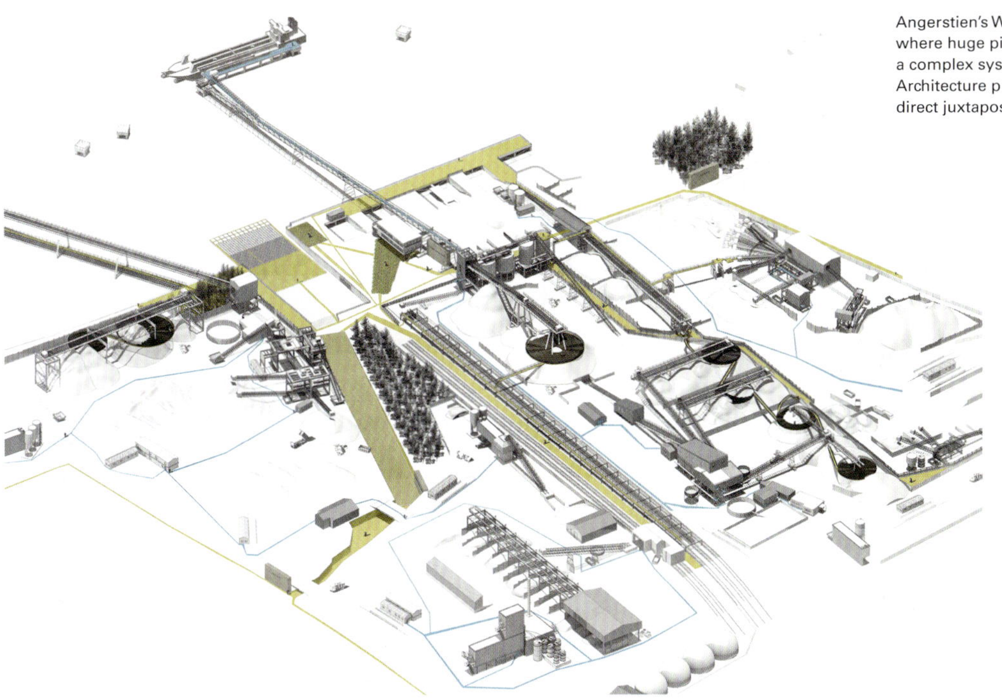

The project creates a new typology of urban relations that allow industry and the public to interact in a way not previously possible. The site purposefully celebrates the flow of materials. Pedestrian spaces are opened and closed by the movement of aggregate routes through the site, evolving in line with industrial cycles, curated levels of engagement and intensities of interaction.

The flows of materials become the basis from which a new network of public spaces and actions are proposed

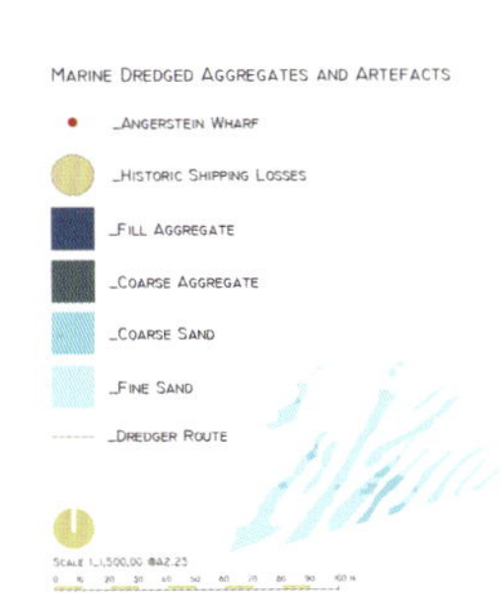

Aggregates are dredged from across the North Sea, connecting Angerstien's Wharf to a wider network of marine landscapes. The wharf is also located in London where metropolitan land values question the future of such a large-footprint industry that provides essential materials to the higher-value urban developments.

Differentiating-Between

If each landscape is considered unique and if we advocate for site-specific combinations of research and design methods to understand and reimagine these places, then an embracing of difference in ways of working can be considered an essential aspect of Advanced Landscape and Urbanism. Simultaneously, to learn from experience, to build on what has been achieved before, and to avoid a homogenisation through diversity has required a clear structure that frames common issues, sites and practices. A differentiated diversity is at the core of Advanced Landscape and Urbanism, across those who teach, research and study, between projects, and across definitions. Each tutor and academic is encouraged to provide a focus based on his or her expertise. The Master's programmes consist of contrasting studios so that a diversity of practices and projects can be fostered within focused design discourses and research agendas. From within a common discourse agreed by the tutors in Advanced Landscape and Urbanism each year – such as exploring relations between 'development' and 'maintenance' as set out by New York artist Mierle Laderman Ukeles in her *Manifesto for Maintenance Art* (1969) – studio tutors interpret and give direction and coordination though design briefs that narrow further, geographically and thematically. There is an intentional distance between the approaches taken by design tutors, from studios focused on 'Landscape Politics' to 'Nothing Architecture' and from 'London National Park City' to '100-Mile City', aiming to highlight the infinite possibilities in practising landscape. The contrasting studios are maintained by the different cultural and disciplinary backgrounds of design tutors whose own training in fields of urban design, architecture, art, literature and philosophy – and of course landscape architecture – informs their approach.

Student projects are pursued in small collaborative studios, where sharing knowledge and participating in critique are essential activities to realise individual designs. Advanced Landscape and Urbanism has no singular style, but through pedagogical invention and an emphasis on specific research and design methods, as well as techniques of advanced representation, we have developed and encouraged several 'ways of working'. These include three composite drawings, termed 'base drawing', 'operational drawing' and 'scene', that each mark significant moments in the design process: firstly, concluding the research with a single drawn foundation from which proposals can be generated (base drawing); secondly, visualising the complex relations that are the basis of the design proposal (operational drawing); and third, an image that communicates the construction of the landscape as it is experienced (scene).[9] Each studio employs these ways of working with base drawings, operational drawings and scenes differently; however, the questions asked are consistent: What are the conclusions from the research as represented in a single drawing? How can the processes of making landscapes, through social, architectural and ecological means, be represented? How can the experience of landscapes be constructed and communicated?

Tensions-Between

From within and between the messy entanglement of landscapes that we perceive, experience and create, there are frequent tensions and contradictions. How can design projects challenge dominant practices of urban development, which favour beautification and gentrification over spatial equality and secure tenure of housing – while at the same time creating visually stunning designs? How

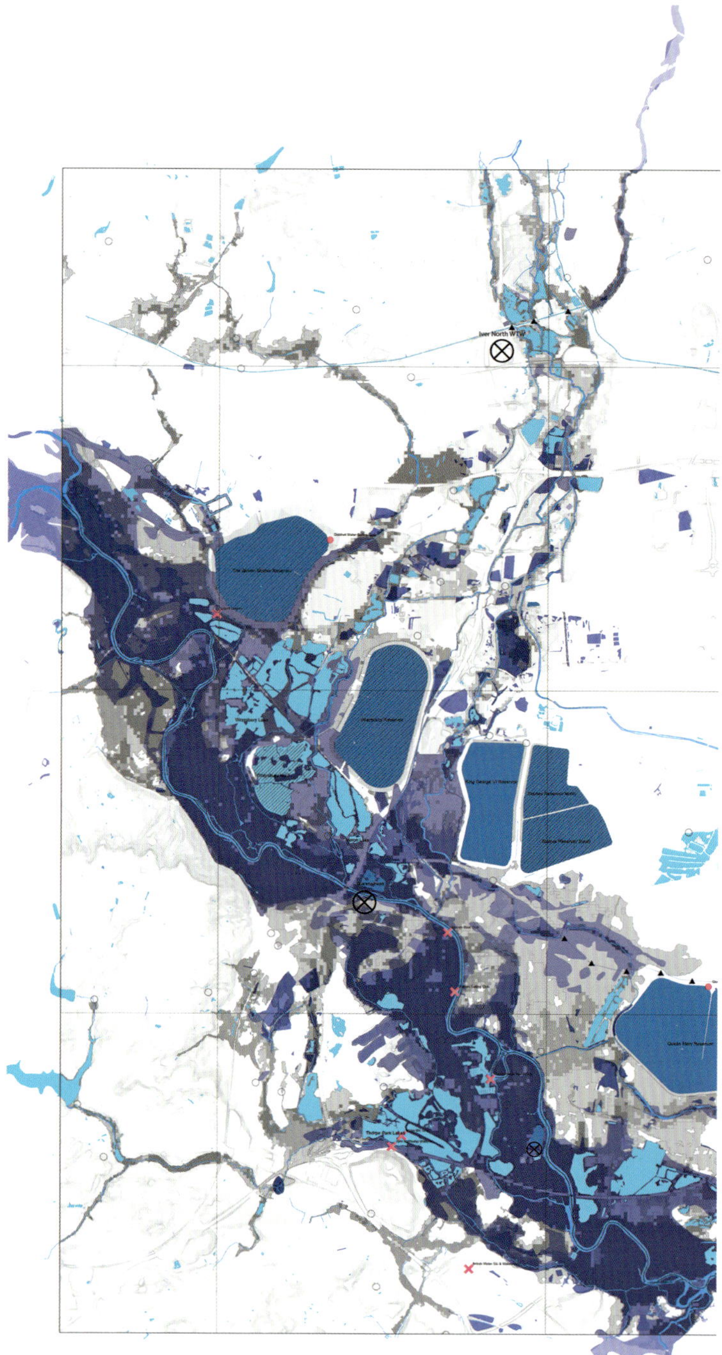

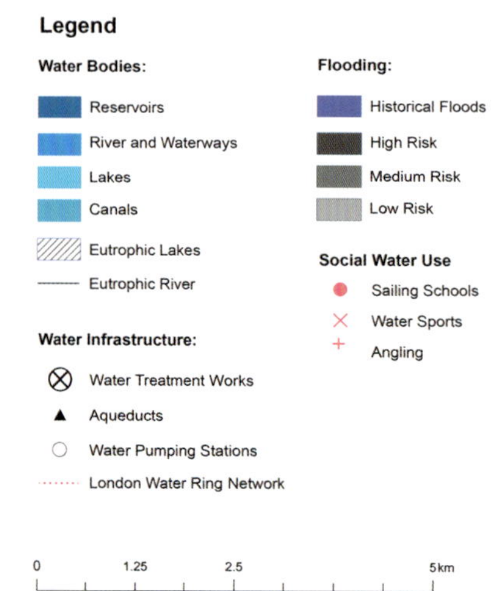

can practices of transforming landscapes and cities address global climate change when the use of carbon-emitting materials such as concrete and the sourcing of building materials from distant countries remain prevalent? As Jane Hutton states: 'The material assemblies of constructed landscapes generate ecological, economic and social conditions in situ, yet are embedded with the relations of their own production, concealed through the processes of commodification.'[10] The search is not for solutions – or even to identify problems – but rather to make some sense of the past while asking questions of the future. By focusing on 'how' landscapes are created, and by 'whom', as a means to produce speculative proposals, many of these contradictions can be explored. MA student Rosie Martin's process of developing proposals for homeless men on the Old Kent Road personifies many of these struggles. She states: 'These moments of process cannot always be seen in the final outcome.'[11]

'Changing the world' through landscape projects, as advocated by postgraduate coordinator Benz Kotzen,[12] requires understanding of the means through which such ambitions can be achieved and the ability to effectively communicate them. These relations of landscapes are not passive, but focused actions directed to establish rigorous research projects and speculative designs. 'What *is* landscape at [the University of] Greenwich?' James Fox asked in 2015: it is, and always will be, defined by the excellence of research, design and teaching pursued by the students, tutors and academics involved, and Advanced Landscape and Urbanism is their collective endeavour. ∞

Notes

1. Rosie Martin, 'Making Process', unpublished Master's project book, Advanced Landscape and Urbanism, University of Greenwich, London, 2018.
2. Quote from group interview/conversation with 12 members of Advanced Landscape and Urbanism, conducted on 20 February 2019, with discussion on the future of landscape education through consideration of projects inside and outside the university.
3. Doreen Massey, 'Landscape as a Provocation: Reflections on Moving Mountains', *Journal of Material Culture*, 11 (1/2), 2006, p 40.
4. Ed Wall and Alex Malaescu (eds), *Testing-Ground: Journal of Landscapes, Cities and Territories*, Advanced Landscape and Urbanism (London), 2016.
5. Jane Hutton, 'Reciprocal Landscapes: Material Portraits in New York City and Elsewhere', *Journal of Landscape Architecture*, 8 (1), 2013, pp 40–47.
6. Caroline Knowles, *Flip-Flop: A Journey Through Globalisation's Backroads*, Pluto (London), 2014.
7. Hutton, *op cit*, p 46.
8. *Ibid*.
9. For more on base drawings, operational drawings and scenes, see Ed Wall, 'What … is Landscape?', in Karsten Jorgenson *et al* (eds), *The Routledge Handbook of Teaching Landscape*, Routledge (Abingdon), 2018.
10. Hutton, *op cit*, p 40.
11. Martin, *op cit*.
12. Group interview/conversation with 12 members of Advanced Landscape and Urbanism, 20 February 2019.

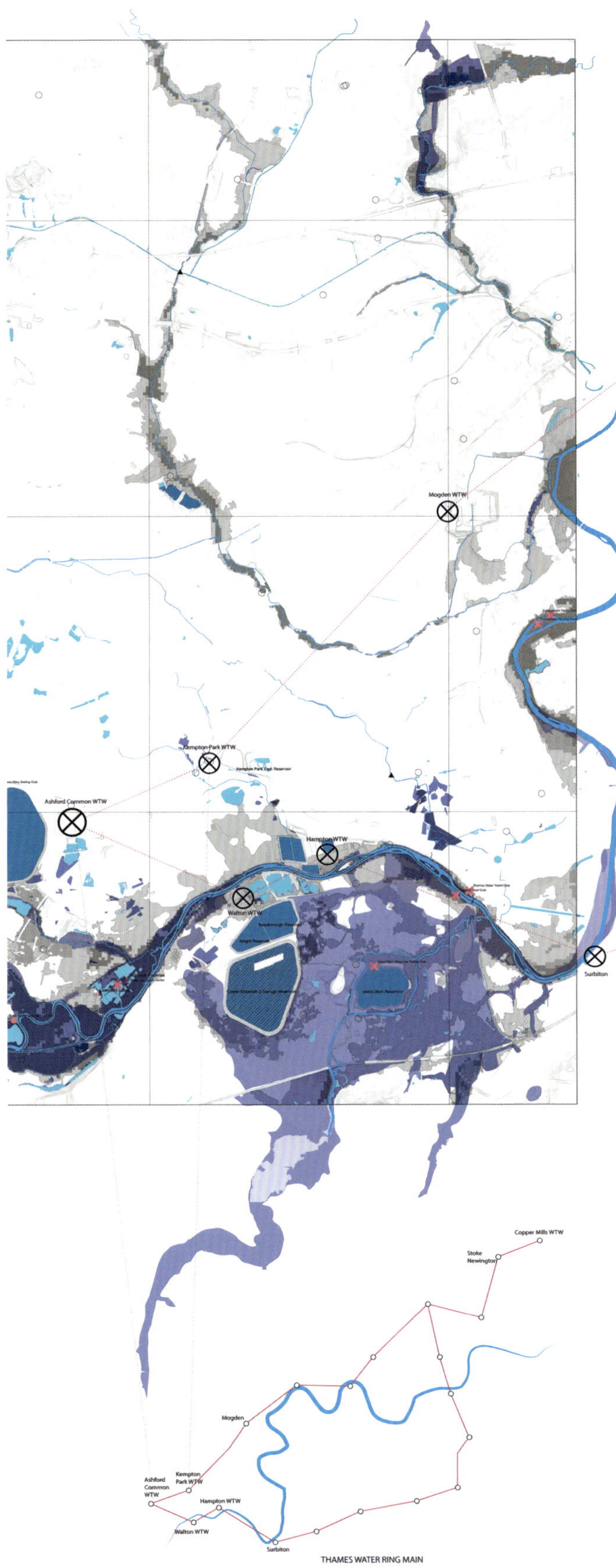

George Armour,
Edgelands,
Advanced Landscape and Urbanism,
University of Greenwich,
London,
2019

Focused base drawings developed by MA Landscape Architecture students provide for unique readings of landscapes, such as of this peripheral urban landscape near London's Heathrow. Mappings of many conditions of wetness provide a unique lens through which to establish bold infrastructural interventions and more subtle, localised actions.

East Anglia Records,
'Album Launch',
Slade School of Fine Art,
University College London (UCL),
London,
2016

Documentation of a party. East Anglia Records started as a party night at the Slade and has gone on to produce music, events, illustrations and performance. Organising a party involves, among other tasks, making a poster, finding audio cables, and asking people to DJ. The practice believes that such roles can be viewed as those of a landscape architect, determining types of exchanges by making amendments to existing conditions.

MEAL-DEAL ECOLOGIES

LANDSCAPE THINKING

Fred Duffield,
Hubcaps,
'Mark II' exhibition,
Woburn Square Garden,
Bloomsbury, London,
2015

For 'Mark II', a group of artists made work in relation to parts of a Nissan Micra. Duffield's piece on the car's hubcaps is an amendment to a familiar scene, a car in a car park, which resets an idea of where it has been and where it might be going.

Harry Bix is the founder of art practice East Anglia Records, and teaches Landscape Architecture at the University of Greenwich. The post-Situationist, event-driven ethos of his work revels in the everyday and incomplete, without seeking fully quantifiable definition. As he explains here, his projects are happenings, bringing together people, music and ambiences – catalysts operating on existing conditions, rearticulating them as landscapes of events.

Roza Horowitz and East Anglia Records,
'We were merely players' exhibition,
SET Alscot Road,
London,
2019

Artist and East Anglia Records house-band member Harriet Rickard sits in front of Dutch artist Horowitz's painting *The Last Days of a Dictator*. East Anglia Records collaborated with Horowitz to recontextualise her show by performing a song called 'boy for sale' while she performed actions from her two day-jobs: cold-calling telephone scripts and making burgers at Burger King.

East Anglia Records is an art practice that began as a party night. It presents momentary experiences and relationships that emerge over time in order to understand how they work in relation to institutional mechanisms. The practice is presented as a record label that produces music, events, illustrations and performance through which artists can imagine themselves in relation to a temporary institution of sorts. Born of landscape architecture education and then working in practice offices, late-night CAD-ing and producing landscape images, it looks towards the core principles of landscape as a way of operating. Within the office environment it has claimed social roles: DJ-ing on Friday afternoons and making staff birthday cards were early subconscious gestures of its thinking. 'Album Launch', a series of music performance nights, came out of a basic social need for a bar at art school.

The practice views landscape as a set of relationships, and landscape architecture as the design of relationships between objects and activities through time – but never complete objects or relating to a specific time. A landscape architect determines types of exchanges by making adjustments to existing conditions. East Anglia Records asks: How might a landscape architect design the landscape in moments and in daily actions? How might an individual's encounters, walking streets and riding buses, present the landscape itself?

Relational Activities

The autofiction *summarynjdkjds.pdf* (2018) recalls a
year of discussions of a failed collaboration between
East Anglia Records, Southwark Park Galleries and the
Millwall Community Trust in London who were organising
a participatory art workshop for alternative-provision
students. In four parts, it documents an individual moving
through a series of spaces and their interactions with
people along the way. Some spaces described are directly
linked to the collaboration, while others are unrelated
experiences that occurred at the same time. By observing
specific details like clothes worn in meetings at Southwark
Park Galleries alongside conversations during a golf trip
in Nottinghamshire, for example, *summarynjdkjds.pdf*
tries to understand where conceptual mechanisms like
masculinity, class and art present themselves.

For the exhibition 'Mark II' (London, 2015), artists were
invited to respond to the existing conditions of a car,
parked in a street. Each was assigned a part of the car:
back seat, stereo, air vents and so on. A mother-of-pearl
house key by artist Cristine Brache hung from the ignition;
hubcaps by artist Fred Duffield were sprayed with pink-
and-white plaster and betting slips were half-attached,
falling on the floor below as if a quick getaway from
another place preceded the moment it was parked; while
around 30 people stood and watched artist Sarah Boulton
(air vents) read a text in a car with all the doors shut,
turning a normal scene (person in a car) into a spectacle.
All of the work relies on the audience's strong associations
with a car: how it works, its temporalities and what it is as
a space, both culturally and socially – an object that passes
through the landscape, stopping from time to time.

You are the memory of the car and you are flashing
back, revisiting every possible thing the car has ever
experienced at once.[1]

Harry Bix and Panicattack
Duo and Friends,
The Sponge,
The Change Room,
Deptford, London,
2019

As part of a collaborative performance
with artists Panicattack Duo and
Friends, East Anglia Records founder
Harry Bix performed an evolving drum
loop that took on new sounds as he
continued throughout the space of a
changing room in a former college in
Deptford. For the album *Enjoy every
moment* (2018), the practice used
drum loops and the change of a loop
over time to illustrate walking and
passing through time.

Harry Bix,
Handwritten notes redrawn
as proposed landscape plans,
Tate Modern,
London, 2019

Edited with Tipp-Ex, this illustration formed part
of the backdrop for the launch of issue 235 of art
and literary magazine *Ambit*. It explores a plan
drawing arrangement by tracing around and
overlaying quickly handwritten notation.

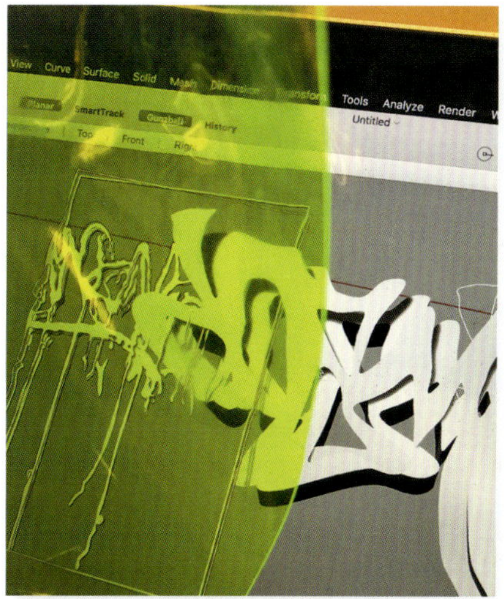

right: A 3D digital model drawn from an image of a notice board found in London and a sheet of luminous acrylic waved in front of the computer screen by artist and house-band member Hannah Le Feuvre. The photograph begins to speculate new forms and arrangements by overlaying two moments of a day.

For the launch of issue 235 of *Ambit* magazine at Tate Modern, London (2019), the East Anglia Records house band performed a tender repeated jingle in between poetry performances by contributors to the magazine. Behind them as a backdrop, panning zoom-ins of hand-drawn speculative landscape plans played on loop. The band includes studio members of SET Alscot Road artist studios, and the sub-story of their performance is the display of their friendships cultivated there, as an energetic product for the night. The house band has provided this service at other art, performance and music events with different sets of members varying in number depending on who might be available and interested at the time.

Released by East Anglia Records in September 2018, the album *Enjoy every moment* documents the summer that year as music. Using a range of recorded studio conversations, the improvised thoughts and intonations of artist Hannah Le Feuvre and looping drum rhythms that feel like the experience of walking, it offers a pseudo-philosophical understanding and translation of a particular period of time.

A Guide of the Land

Like a landscape, these exchanges do not seek definition. Maybe the greatest strength of the landscape is that it exists outside of an image of itself. Individually we are vessels of the landscape, and a landscape architect may be the entity most aware of that.

> …chewing gum sealed table, tears in my lasagne, stain on my jumper. - 'hey, how's it going, what you been up to' architecture, 'yeah, I'm alright' architecture. Lea, Sara, Becky and I go around as we do.[2]

East Anglia Records suggests that a future landscape architect may operate as a guide of the land, walking about, moving objects and introducing people to one another. ᴅ

Notes
1. Cristine Brache (formally referred to as Audrey Phillips), 'Stalking *Mark II*', AQNB, 3 June 2015: www.aqnb.com/2015/06/03/stalking-mark-ii//
2. Harry Bix, quote from 'Meal Deal Ecologies' text originally performed at the exhibition 'you're mulchy green, you're verdant matter' at the Slade School of Fine Art, University College London (UCL), 2019

WORKING PLACE

CONSTRUCTING COLLAGE AS CRITIQUE

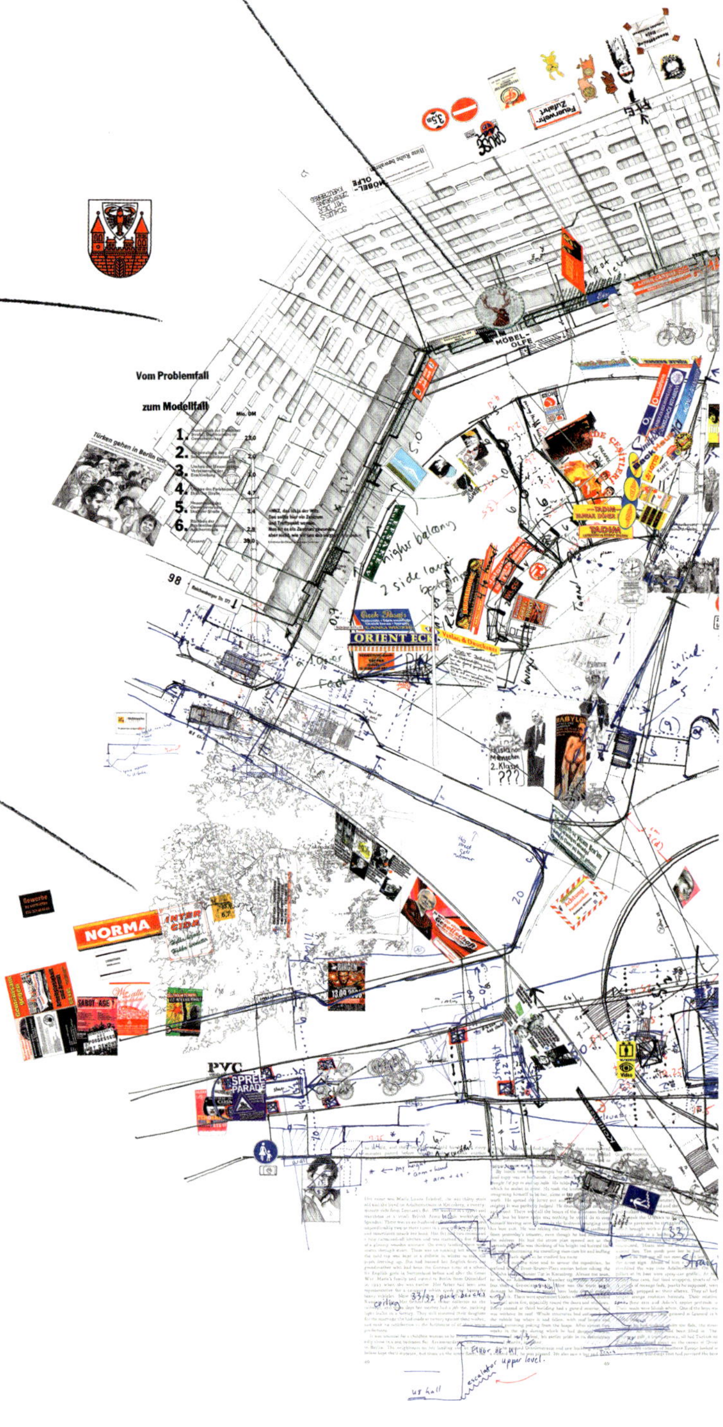

Larissa Fassler,
Kotti, 2008,
Kottbusser Tor,
Kreuzberg, Berlin,
2008

Collage depicting place and people in 2008 in Kottbusser Tor (nicknamed 'Kotti') with the Neue Kreuzberger Zentrum (NKZ) building as the backdrop.

Landscape architect **Toya Peal** explores the work of Berlin-based artist Larissa Fassler in a quest for new ways to represent civic, social and urban space. She highlights the potential role of Fassler's social mappings, drawings constructed from detailed observations that provide multiple perspectives of specific sites, to critically inform planning and design processes.

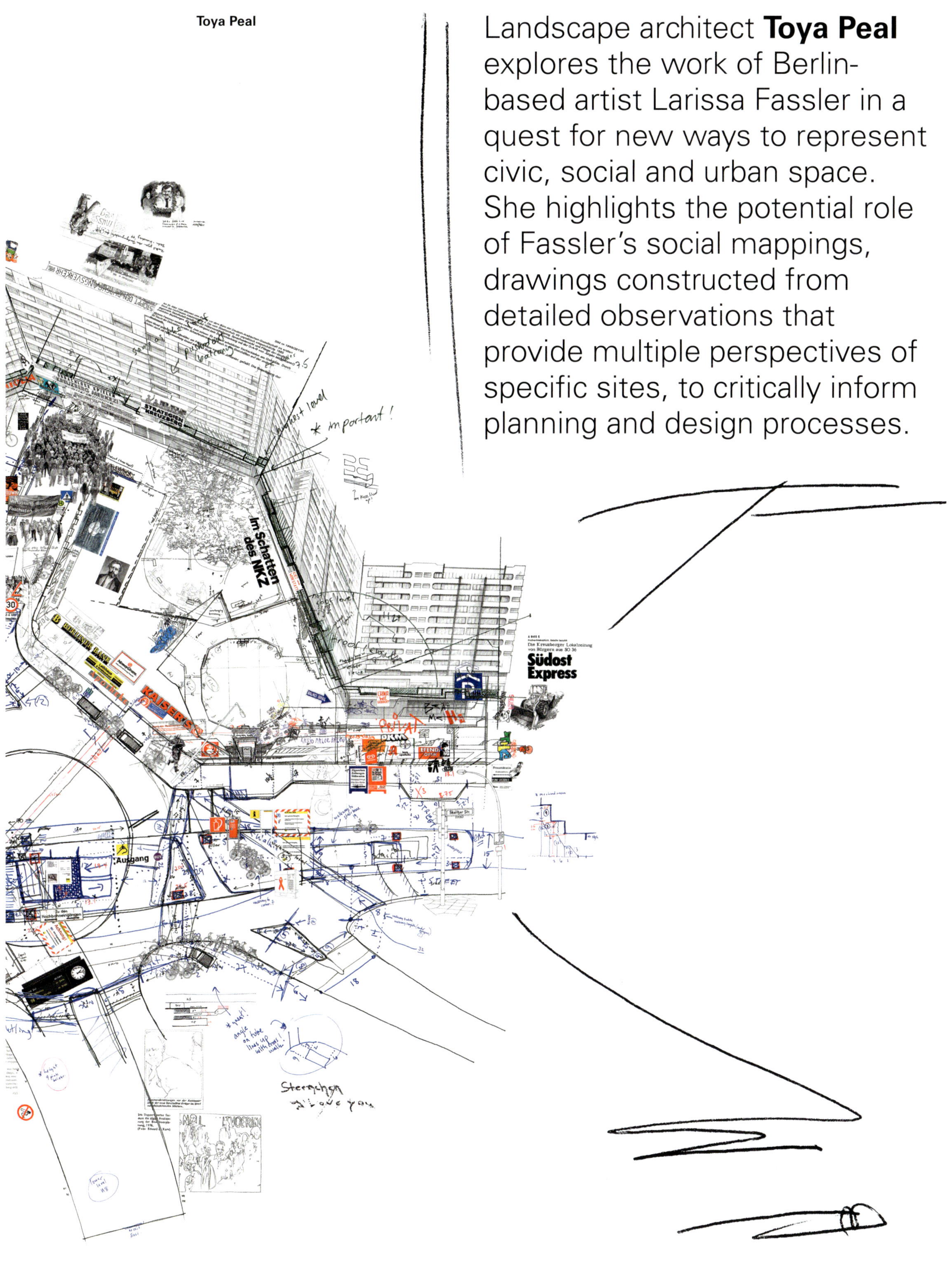

A space must be taken over in which men construct, perform, entertain, experiment.

— Derek Jarman, 2011[1]

Larissa Fassler, a Canadian artist living in Berlin, observes, documents and draws public spaces of cities as part of her artistic practice to understand and portray the relationship between a place and the people that occupy it. In 2004 she began studying public places in Berlin. Using architectural methods of representation to capture the place, her quotidian surveillance and investigation into the operation of the space is collated and overlaid like a collage.

The mixed-use redevelopment block Neue Kreuzberger Zentrum (NKZ) forms the backdrop to Fassler's collages in the triptych 'Kotti'. The building, designed by architects Wolfgang Jokisch and Johannes Uhl, was completed in 1974 as part of wider postwar redevelopment plans in West Berlin. It is a landmark for Kottbusser Tor, the U-Bahn station and station square (fondly nicknamed 'Kotti') over which it towers, enclosing the public space. Straddling the main road through Kreuzberg, the NKZ block is an amalgamation of stacked housing and transport interchanges as it also connects to the subway, bearing similarities to buildings typical of the radical urban planning of the 19th century.

Photomontage by artist Larissa Fassler, where the NKZ building stands as a backdrop to Kottbusser Tor station square. The mixed-use development provides a plethora of activity that animates the public space.

The three collages *Kotti, 2008, 2010* and *2014* each took approximately a year to complete. They present like an amphitheatre with different stage sets, yet the enclosing NKZ building provides a constant scene. Fassler's variety of detailed handwritten observations, annotations, sketches and research are cut out, placed and overlaid like a collage. These observational materials animate the stage sets, and despite the chaotic appearance the overlays suggest a rhythm to the place. Are these collages the stage script, or palimpsest? Perhaps operational drawing, or even social mappings?

Although the 'collages' are large and near-human height, the layers of physical and subjective detail of the place embedded within the images draw the viewer in to the point where one's nose is nearly touching the image. Looking closely through the artists' magnifying glass, the work focuses on the relationships between places and people. The physical qualities of the spaces are recorded in footsteps and body heights, overlaid with associated research, newspaper pages, stories, historical events, advertising signs and slogans, all of which are redrawn by Fassler. Everything we see in her work has been collected and filtered through the artist to understand and make visible how these spaces impact people both psychologically and physically.

Should this methodology of social mappings be shared with urbanists and planners in order to value and provide critical interrogation of urban civic spaces as a design tool to develop places for all people?

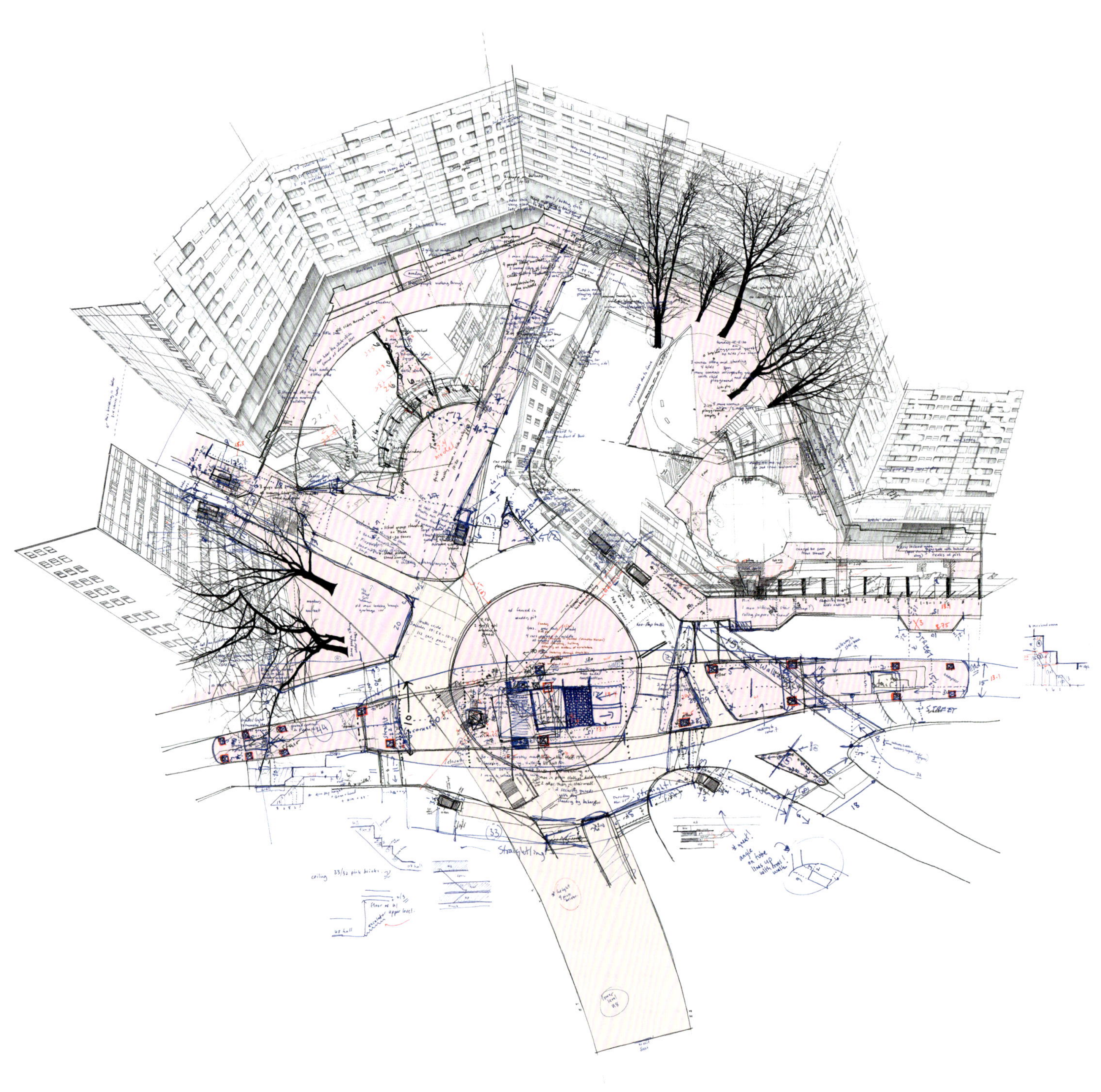

opposite: The collage shows what is public and private in Kottbusser Tor. Coloured areas depict public space, and white areas private space. The NKZ building remains a constant backdrop.

below: Fassler's collected observations of advertising slogans, signs and social messages are collected and reproduced by hand drawing and then scanning in to create a collaged composition of space and people in Kottbusser Tor. The NKZ building remains as the backdrop.

Collage as Theatre

If interpreting Fassler's 'Kotti' series as an amphitheatre, who is the audience? Who is the stage director and what does this story of the space tell us? What does this seemingly chaotic social representation of Kotti reveal about the success or failure of its public space? Should this methodology of social mappings be shared with urbanists and planners in order to value and provide critical interrogation of urban civic spaces as a design tool to develop places for all people?

Despite its dark passageways, the dominance of NKZ's social housing and negative associations with crime and violence, Kottbusser Tor is simultaneously a thriving urban civic place offering social spaces for a conglomeration of people at all times of the day. Punks, LGBT communities, tourists, students and Turkish immigrants occupy its spaces 24/7; markets and cafes by day, nightclubs and bars by night.

The artist frames *Kotti, 2008, 2010* and *2014* from the same imagined viewpoint from above, but by doing so is able to capture the whole scene in one image. Fassler spends hours, days and months observing, photographing and revisiting the space to record information to construct her collages

Layered signs and observations create a dense and lively streetscape to animate the space in front of the NKZ building.

and establish her physical and critical position on the public spaces of cities. Particularly in *2008* and *2014,* the 'collaged' annotations, text and coloured hand-drawn images keep the viewer's eye constantly moving around the image and thus provide many aspects. The perspective is distorted, and the sense of time removed; the handwritten observations exist all at the same time in a kind of organised chaos. As a result, Fassler's collages offer a pan-optical view of Kotti, an alternative to the traditional eye-level scenographic views associated with evolutions of landscape.

Hyper-activity

Landscape was once termed 'a cultural image, a pictorial way of representing or symbolising surroundings',[2] yet this definition seemingly only applied to when landscape was purely the subject of painting in the 18th century. Now landscape is complex, layered, lived – and cities too. In the mid-20th century, Jane Jacobs appealed against the architectural styles focusing on the appearance and configurations of cities, instead advocating the importance of how a city operates. She first argued that all aspects of a city – its streets, parks and dwellings – should be diverse in use to 'give each other constant mutual support, both economically and socially'.[3] Secondly, that cities be dense, creating liveliness and hyperactivity. Together, density and diversity may stimulate unusual encounters, discovery, a sense of place – even comedy. Through Fassler's observations and collages, people come together both in reality and through imagination.

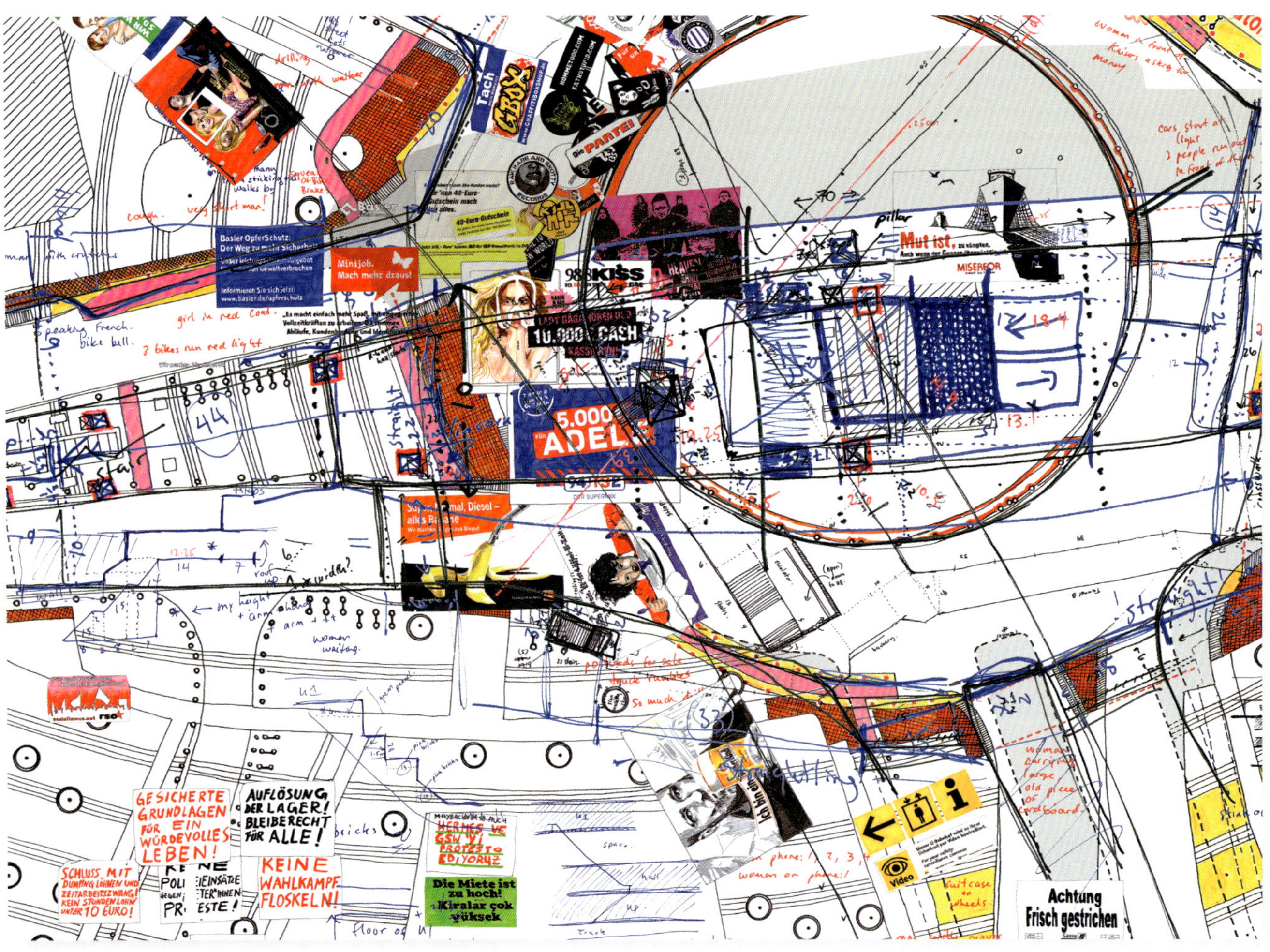

In *Kotti, 2008*, for example, an excerpt from an Ian McEwan novel, handwritten by Fassler, depicts a fictional character who lived in Kotti; below is a pencil-drawn portrait of a man wearing a sandwich board reading 'Are foreigners second class citizens?'

What does density and diversity bring to Kotti? It brings artists. William Empson famously quoted that 'the arts result from over-crowding'.[4] However, artists bring the bourgeois. Capitalism has taken interest in Kotti, with rental prices increasing to the point of toppling the diversity and social balance. Jacobs alludes to this issue as the 'self-destruction of diversity',[5] whereby the popularity of a neighbourhood with such vitality can become profitable. This is a recurring issue in Berlin, and consequently has become a part of Fassler's focus, for example in her more recent project *Emotional Blackmail* (2018). Fassler was commissioned by the KW Institute for Contemporary Art to create temporary artworks on billboards to highlight social issues arising from planned high-end developments in the Moritzplatz neighbourhood by real-estate firm Pandion. Here, her applied method of pan-optical observations of people and place helped her to create billboards that stand like mirrors, offering a true reflection of Moritzplatz.

Will developers take notice of the social issues and Fassler's foreseen consequences? Hard to tell. But nevertheless her work captures a mood, a reality, and the voice of the people who live in the neighbourhood, and this could be an important catalyst for change.

Perhaps evaluations like Fassler's animated triptych should be used by designers and 'city operators' to critically interrogate urban civic spaces. An ethnographic study of people and place, the work conveys both the comical and the dysfunction of Kotti, which overall suggests its sentimental value and success as a diverse cultural space. The palimpsest script suggests a vulnerable but much-loved character. ⌀

Public billboards by Fassler stand in the Mortizplatz neighbourhood as part of the REALTY project commissioned by the KW Institute for Contemporary Art. The boards reflect back to residents and developers the social issues arising from proposed developments by real-estate firm Pandion.

The private and public spaces are represented as positive and negative forms in Fassler's architectural model of the NKZ building. Approximate scale: 1 footstep = 3 centimetres (1-inch).

Notes
1. Derek Jarman, 'Tentative Ideas for a Manifesto After 1 1/3 Years at an Art School', in Alex Danchev, *100 Artists' Manifestos: From the Futurists to the Stuckists*, Penguin (London), 2011, p 374.
2. Kenneth Olwig, 'Representation and Alienation in the Political Land-scape', *Cultural Geographies*, 12 (1), 2005, p 19.
3. Jane Jacobs, *The Death and Life of Great American Cities*, Vintage (New York), 1961, p 14.
4. Richard Sennett, 'The Open City', in Tigran Haas and Hans Westlund (eds), *In The Post-Urban World,* Routledge (London), 2017, p 100: richardsennett.com/site/senn/UploadedResources/The%20Open%20City.pdf.
5. Jacobs, *op cit*, p 243.

A snapshot in time showing the *Sky Garden* going about its Surrealist mixing
motions, this time evoking paint splatters to enliven the mix.

Neil Spiller

LANDSCAPE DRIFT

SOMETHING IN THE AIR TONIGHT

How can contemporary technologies conjure up new landscape forms that not only create new typologies of landscape, but also new methodologies of space–making? Architect and Editor of ◿ Neil Spiller illustrates one of his attempts to transcribe these notions within the context of the *Sky Garden*, part of his magnum opus Communicating Vessels project.

Even within the most beautiful landscape, in trees, under the leaves the insects are eating each other; violence is part of life.
— Francis Bacon, 'I painted to be loved', 1992[1]

(A)rchitecture is a movement beyond material. It is length, height, and width, but also the depth of aspiration and memory.
— Daniel Libeskind, *The Space of Encounter*, 2002[2]

The garden has been commonly used as an analogy for virtually everything: life, mental and physical wellbeing, promiscuity and even for hallucinogenic drug experiences. This all, perhaps, stems (no pun intended) from the mythical Garden of Eden. Eden, at first a paradise of human innocence, a feast of serene natural beauty, then latterly the Arcadian backdrop to human sexual awakening, shame, desire and guilt. So the garden is a Rebis thing, an alchemical, double-sided concept simultaneously containing its opposites – violence and love, clash and desire.

THE CITY AS A GARDEN
Likewise, the city is a garden, a surreal garden. For the Surrealists, their searching in the city of Paris during the 1920s was for the same phenomena they found so appealing in the paintings of Giorgio de Chirico: the extreme juxtapositions of objects, their scale and combination of aesthetics, chance occurrence, impossible vistas (both mentally and physically) and desire creating uncanny situations. The city was seen as a delirious lover and a hallucinogenic space of reflection, poiesis and semiotics, providing a cornucopia of new images and concepts. The City Garden was complicit in the making of all aspects of Surrealism.

As the contemporary landscape architect and academic Fernando Magallanes has written, for the Surrealists the 'Parc des Buttes-Chaumont was a site filled with abstract fictive possibilities and more concrete visible objects, such as oddly placed Greek follies, engineered bridges, and reconfigured artificial landscapes containing magical and psychoanalytical meanings.'[3]

The City Garden is likewise an alchemic alembic vessel, a site of distillation, of cyclic iterations, of putrefaction and rebirth and a search for the ultimate substance – the philosophers' stone. The stone signifies the attainment of some inner knowledge, a panacea, and provides life-giving magic. The Surrealists were big fans of the Alchemists' allegorical writings, their Rebis-type illustrations that accompanied their manuscripts and their arcane rituals. The conceptual alignment between the work of the adept Alchemists and the Surrealist magi, and each other's hungry search for some sort of inner knowledge, did not escape them.

What happens when we bring this experimental dynamism into the 21st century? The *Sky Garden* (2014–16) is a virtual garden that forms part of the magnum opus Communicating Vessels architectural/landscape project begun in 1998. A meta-project, Communicating Vessels seeks to document and explore potential changes between landscape and architecture caused by recent technology, particularly virtuality and biotechnology, using as a microcosm and testing the age-old relationship between house and garden. The project is situated on an island, a magical island.

The island is a landscape of extraordinary spaces, nano-technological fluids, reflexive actions and reactions. The biomechanical 'putty' that holds it all together, the island's liquid lifeblood, is the 'grease' or 'holy gasoline'. Objects are flung from temples of repose and caught on the revolving tableaux of the Dance of Death; there are wayward vistas, twisted christs, singed wedding dresses, genetic gazebos, wobbling wheelbarrows and miniature alchemic laboratories among many other wild interventions. The island is inhabited by a professor of surrealism and numerous numinous presences created by augmented reality. It is bigger than itself; it also has virtual tendrils that stretch across the world, into the water and up into the sky. Augmented and full-blown virtual reality figures large in the work.

GARDEN OF FORKING PATHS
The Communicating Vessels project has a similar anatomy to Jorge Luis Borges's famous short story 'The Garden of Forking Paths' (1941). It is a series of multiple, bifurcating inputs and outputs choreographed by chance, determining trajectories, vectors, actions and reactions across the virtual and actual and the myriad of mixed and augmented spaces between. In his story, Borges writes: 'The Garden of Forking Paths (GoFP) is an incomplete but not false image of the universe'.[4] The Vessels is likewise an attempt, in a microcosm, to illustrate the importance of chance in defining our lives and surroundings, but it also evokes Surrealist attempts to guide an exploration of the spatial opportunities the 21st century gives to architects and landscapists. If we return to Borges, the GoFP is not about dwelling in a uniform, absolute time, but an infinite series of times, in a fecund, vertiginous net of forever entwining and untangling events: 'This network of times which approached one another, forked, broke off or were unaware of one another for centuries, embraces *all* possibilities of time.'[5] This revealing of *all* possibilities of reality and the other partial lives of objects is true for the Surrealist too.

IN AN ENGLISH COUNTRY GARDEN
From 2012, triggered by the passing of friend and experimental architect Lebbeus Woods, the Communicating Vessels island gained a *Walled Garden*.[6] The garden uses a particular point of view of an observer to trigger an augmented-reality storm (Lebbeus died the night Hurricane Sandy hit New York). It therefore has two components, virtual and real, each cosseting and reacting to the other. Of course, the augmented component could be anything, and with this realisation the last drawing of the series shows a golden storm.

Shortly after the *Walled Garden* drawings were created, during a visit to Berlin to talk about and show them at the opening of the 'Lebbeus Woods ON-line' exhibition at the Tchoban Foundation in 2014, a visit to the Altes Museum fuelled an already present interest in the work of architect and painter Karl Friedrich Schinkel (1781–1841). His work at the Charlottenhof Palace, Potsdam, for the Crown Prince of Prussia proved fascinating, but particularly his garden plans and use of the ancient hippodrome form in the landscape inspired by Pliny the Younger's description of his Tuscan Villa in the 1st century AD:

The design and beauty of the buildings are greatly surpassed by the riding-ground [hippodrome] … At the upper end of the course is a curved dining seat of white marble, shaded by a vine trained over four slender pillars of Carystian marble. Water gushes out through pipes from under the seat as if pressed out by the weight of people sitting there, is caught in a stone cistern and then held in a finely-worked marble basin which is regulated by the hidden devise so as to remain full without overflowing. The preliminaries and main dishes for dinner are placed on the edge of the basin, while the lighter ones float about in vessels shaped like birds or little boats.'[7]

The seating form Pliny describes is called a stibadium. Schinkel's hippodrome also contains a stibadium at its curved end. Likewise, Pliny's description of the hippodrome and stibadium with its floating bowls of food were the inspiration for the Communicating Vessels *Sky Garden*. It needed to have a spatial relationship with the *Walled Garden for Lebbeus*, to exploit new technology and have the potential to operate as an analogue computer, mixing and hybridising ideas – working as a Surrealist chunking engine, provoking new notions of space and form.

left: The *Golden Storm* is the augmented-reality component of the garden, activated by the viewer placing their face and ears within the hollowed-out head of the statue of Electra seen bottom right.

The *Sky Garden* is positioned above the *Walled Garden for Lebbeus*, which is often seen through it. They are connected in this way.

ASTRAL BODIES AND THE VERTIGO OF THE VIRTUAL

While the form of the *Sky Garden* is a traditional hippodrome with a stibadium at its curved end, it was initially site-less and its surreal purpose unclear. Surrealism revels in reconciling opposites (the fluid and the rigid, for example), conjoining disparate objects and concepts (such as weather in interiors). The use, aesthetic effect and site and motion of the *Sky Garden* were elaborated by drawing, by being open to chance and then rereading the drawings produced – an attitude to architectural production that in some sense dislocates the trained architectural self, liberating ideas that are perhaps alien to traditional architectural thought patterns. The first drawing mapped out the overall form of the garden, but strangely featured an angel, reading a book, incongruously perched on the side of the plan. Why? When reread the angel implies an otherworldly position, perhaps somewhere but not there, existing in a virtual reality as angels do.

The Sky Garden does not exist in the normal sense; it is fully virtual, accessed only through virtual-reality goggles. Once donned, the goggles reveal a surreal garden that has no ground. It is also a vertigo garden, high above the Communicating Vessels island. The second drawing suggested it surfing on clouds.

Neil Spiller,
Sky Garden,
Communicating Vessels,
2014-16

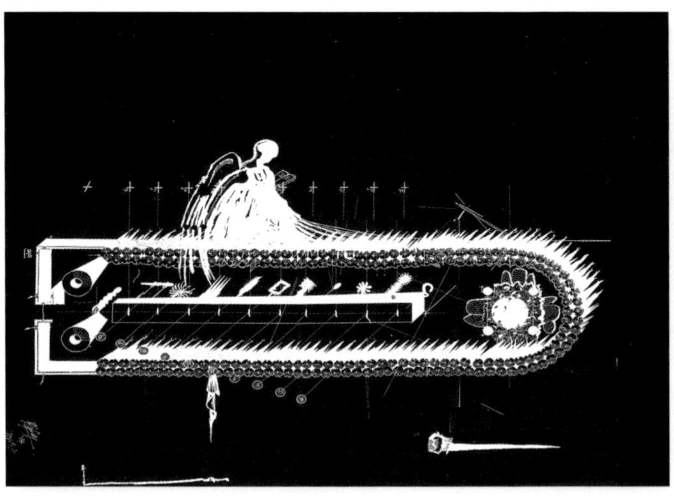

The initial drawing of the *Sky Garden* clearly shows its *spina* and the symbolic statues mounted on it, as well as its incongruous angel. The *Sky Garden* surfs on clouds. Its own virtual stibadium can be seen at the rounded end of the hippodrome.

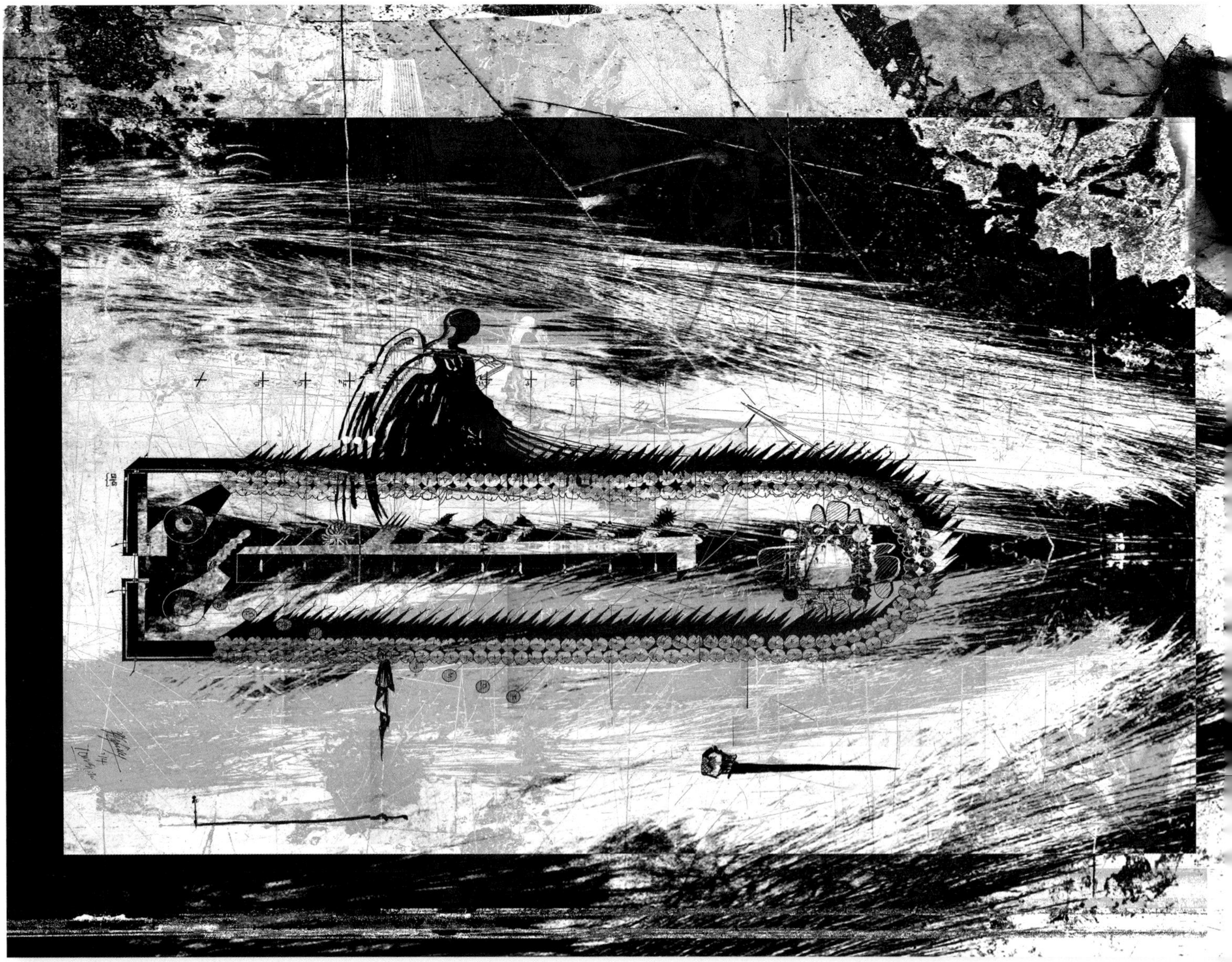

The initial drawing of the *Sky Garden* with notional vectors applied.

The virtual *Sky Garden* can reconfigure itself in an infinite number of ways to encourage inspirational juxtapositions of objects and vectors. The landscape can be seen below.

It constructs a mobile, inspirational landscape of possibilities

In the ancient world, hippodromes had symbolic functions. In Constantinople, for example, the centre of the Hippodrome, the *spina*, was decorated with the artistic spoils of empire. These included obelisks such as the Tripod of Plataea, built to commemorate the Greeks who defeated the Persian Empire in the 5th century BC, but removed from the Temple of Apollo at Delphi by Constantine the Great in 324 BC and set in the middle of the Hippodrome. So what was set on the *spina* was of great significance to a city's social order and its representation of itself and its citizens. The *Sky Garden* hippodrome also has a *spina* bedecked with artefacts of great significance.

A series of forms, each representing some of the major elements of the Communicating Vessels project – the *Meat Hook*, the *Baguette*, the *Holey Hedge* and the *Clinamen* – are arranged at the hippodrome's centre. All of these emblems are vital – their presence in the *Sky Garden* allows it to function as a Surrealist mixing engine.

The workings of this kaleidoscope of spatial and formal interactions became clearer as a series of loci were added to the drawing. Virtual, orbiting, weightless objects (based on Surrealist history, for example, a sewing machine, an umbrella, door or a soft watch) pass through the objects on the *spina*, which might also become detached and join this great combinatorial waltz of inspiration. The professor, inhabitant of the island, far below, is transported by virtual reality into the ghostly, rotating gearage of the *Sky Garden*'s Surrealist 'difference engine' as it computes by reconciliation and hybridisation to form a beautiful garden of forking paths in time and space. It constructs a mobile, inspirational landscape of possibilities, dislocating the professor's creative self, liberating him from his own artistic preoccupations and egging him on to push the envelope of his imagination – a modern-day exquisite corpse constantly reconfiguring. ᗭ

Notes
1. 'Francis Bacon: I painted to be loved', interview by Francis Giacobetti conducted in February 1992, published in *The Art Newspaper*, 137, June 2003, pp 28–9.
2. Daniel Libeskind, *The Space of Encounter*, Thames & Hudson (London), 2001, p 73.
3. Fernando Magallanes, 'Landscape Surrealism', in Thomas Mical (ed), *Surrealism and Architecture*, Routledge (London and New York), 2005, p 222.
4. Jorge Luis Borges, 'The Garden of Forking Paths', *Labyrinths*, Penguin (London), 2000, p 53. Originally published 1941.
5. *Ibid*.
6. See Neil Spiller, 'Detailing the Walled Garden for Lebbeus', in Mark Garcia, ᗭ *Future Details of Architecture*, July/August (no 4), 2014, pp 118–27.
7. Pliny, *Letters*, book 5, letter 6, 1st century AD, trans JB Firth (1900).

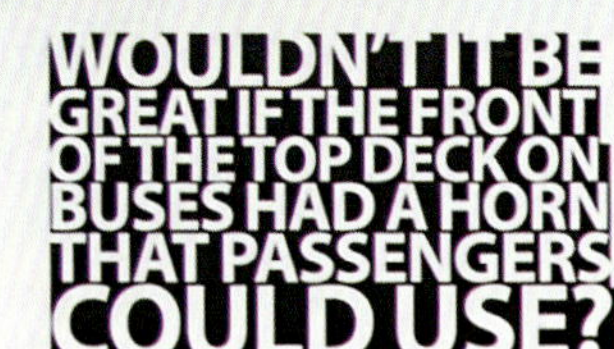

TIME PORTALS, LOVE MACHINES, LAND ORACLES

Thomson & Craighead,
London Wall,
'Being Social',
Furtherfield Gallery,
London,
2012

As far back as 2012 work at Furtherfield was anticipating the interpenetration of the physical and the digital. In this work, Thomson & Craighead present the manifestations of the sensate and emotional city, using publicly available status updates from websites like Twitter and Facebook. Then a form of concrete poetry was assembled from this onto the wall of the gallery.

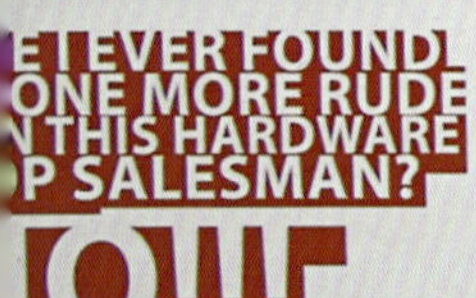

HYBRID GEOGRAPHY AND THE SITUATED DIGITAL

London-based digital art collective Furtherfield creates events and space that encourage cultural flowering through the celebration of difference and otherness. Executive director **Tim Waterman**, a senior lecturer in landscape history and theory at the Bartlett School of Architecture, University College London, takes us on a journey, starting with the organisation's inception, then projecting us into the future with an outline of its projects for the next few years.

'Death Collapsing Into Life' was the title for a walk I led in July 2014 as part of the activist digital art collective Furtherfield's programme to accompany the exhibition of the *SEFT-1 Abandoned Railways Exploration Probe* in their gallery in London's Finsbury Park. The walk followed the route of the Parkland Walk, itself an abandoned railway and now a nature reserve with a pathway that links Finsbury Park with Highgate, not just for walking humans but a dark corridor for teeming bats living in the old brick tunnels, among many other urban species. Applying notions of ecology alongside a bit of Lefebvrian Promethean utopianism, the walk sought to situate ideas of dying, decomposing, fertilising, making, being and becoming in infrastructures, technologies, society, culture and nature.

That same summer I, along with Furtherfield co-founders Ruth Catlow and Marc Garrett, inaugurated a reading group called 'Reading the Commons'[1] that involves artists and scholars from a range of backgrounds including law, landscape architecture and Wikipedia to explore similar ideas of how the commons flower, die and become again. Here, though, the concern was in particular with mapping ideas of the digital commons onto the physical and historical commons. The process confirmed that such a project was both difficult and vitally important in a future in which the virtual and physical worlds will be ever more embroiled. Further, it revealed that models of social ecology, such as those developed by the influential anarchist theorist Murray Bookchin, and of posthumanism in the spirit of Donna Haraway – whose feminist work addresses human and cyborg consciousness – and philosopher Rosi Braidotti, are powerful tools to explore nature–society relations, technology–society relations and place–landscape relations. Digital and landscape – place-based – practices have become inextricably entangled, which led us to define this mutually constituted VR/material embeddedness as the 'situated digital'.

Commissioned by The Arts Catalyst and presented in partnership with Furtherfield Gallery, the project took the artists across Mexico and Ecuador in the Sonda de Exploración Ferroviaria Tripulada (SEFT-1), a remarkable road/rail vehicle of their own fashioning. They gathered stories and testimonials along the routes of disused railways. They shared their journeys online at http://www.seft1.com, where audiences were able to track the probe and view maps, images and videos.

'Death needs time for what it kills to grow in,' wrote William S Burroughs in *Ah Pook is Here* (John Calder and Viking Penguin, 1979). True, but so is its opposite: life needs what death kills to grow in; hope has its roots in ruin. This walk, based in the theories of Henri Lefebvre and conducted along a disused railway line, explored how vicious nature and the ruthless engines of urban decay can show a new ecological, situated, embedded way of imagining utopian landscape and digital relations.

The reading group's explorations into this entanglement led into realms of what Owain Jones and Paul Cloke in their book *Tree Cultures* (2002) call 'hybrid geography',[2] a concept they have adapted from the work of Sarah Whatmore. All are invested in creating alliances with others across multiple platforms to acknowledge and value the nonhuman and the more-than-human in our worlds. This is, according to Whatmore, 'concerned with the spaces of social life, relational configurations spun between the capacities and effects of organic beings, technological devices and discursive codes within which people are differently and plurally articulated'.[3]

Such different and plural framings have increasingly become important for Furtherfield's wider mission, which has from its inception been focused upon social change, but has become enriched by theories of embodiment and embeddedness that insist upon the situation of sociality in substantive, lived landscapes. The commons as another organising frame occupies, for Furtherfield, the centre of a triad of social life constructed by Whatmore: that of 'hybridity, collectivity and corporeality'.[4]

The summer of 2014 was important for Furtherfield, as it was still in the process of adapting to new spaces: two separate buildings in Finsbury Park – one a small gallery, opened in 2011, and the other a meeting space with a kitchen, opened in 2012, which we would decide to call 'Furtherfield Commons'. A whole framework for upcoming activities began to gel at that point, a framework which would become the initiative 'Platforming Finsbury Park': this was recognition that Furtherfield's work extended beyond the internet and its buildings into the park itself and the surrounding neighbourhoods. The milieu of Finsbury Park helped Furtherfield connect with many more and more varied people through artworks that invited them, as Marc Garrett says, 'to question the innocence of the devices in their pockets and the intentions of the private corporations that run our online social spaces'.[5] Then it was natural to ask the same sorts of questions of the park itself and the human and more-than-human relations held within and around it.

Platforming Finsbury Park

Furtherfield started developing the concept of Platforming Finsbury Park in 2017 as a response to extensive research – of which 'Reading the Commons' was part – into the unique issues surrounding its urban and social ecology: its hybrid geography. Finsbury Park, one of London's great Victorian-era parks, is at the boundary of three London boroughs (administrative districts), and has historically served as a recreational space for lower-income people. The park was conceived and established after agitation from the area's working class. Then as now it has been a contentious space: in the 1800s due to upper-class concerns about the appropriateness of the provision of a grand park for the working poor; and now about how to share communication and maintenance between the three boroughs in a time of relentless pressures for the slashing of budgets due to the ideology of austerity, and for increasing privatisation of park use, such as for festivals, during the warm season in which public demand for free access is highest.

Through consultation with park users and stakeholders, artists, techies, researchers, policymakers and other local arts organisations, Furtherfield devised an approach to their programming that would focus on developing the cultural value of the park, in the interests of all its diverse users and life-forms, via digital art activities that centre around placemaking and landscape.

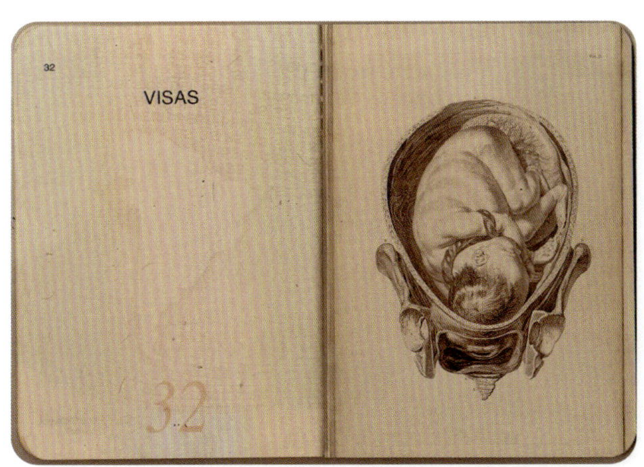

Daniela Ortiz,
Jus Sanguinis,
'Transnationalisms',
Furtherfield Gallery,
London,
2016

Jus sanguinis – 'the right of the blood' – is the sole arbiter for judging the right to nationality in Spain. Living there but of Peruvian descent, Daniela Ortiz's child would thus not have access to Spanish citizenship. *Jus Sanguinis* was a performance in which Ortiz, then four months pregnant, received a blood transfusion from a Spanish citizen, challenging the inherent racism of the 'right of the blood'. The landscape question here is whether dwelling in a substantive landscape should itself be the key to citizenship.

Paul Seidler, Paul Kolling and Max Hampshire,
Terra0, 2016, shown in 'New World Order',
Furtherfield Gallery,
London,
2017

Featured in an exhibition dedicated to exploring art and the Blockchain, *Terra0* is a prototype for a self-owning, self-exploiting forest which can capitalise itself by selling licences for its logging and use of other assets. Built on the Ethereum network, it 'aims to create technologically-augmented ecosystems that are more resilient, and able to act within a predetermined set of rules in the economic sphere as agents in their own right' (terra0.org).

Platforming Finsbury Park has become not just a frame for understanding Furtherfield's future directions, but also a useful lens for casting back to look at past work, key examples of which are illustrated here. This work in total examines forms, modes and products of world-making (or worlding), nature–society relations, technology–society relations and place–landscape relations. Relations may also be spoken of as relation*ships*, and the powerful little suffix '-ship' has been extremely useful to Furtherfield as a tool for thinking. Kenneth R Olwig's philological examination *The Meanings of Landscape* (2019) has been foundational. Starting by showing how the suffixes '-schaft' and '-ship' are cognate, and that they describe a relation, constitution or condition, he goes on to say: '*Schaft* is related to the verb *schaffen*, to create or shape, so *ship* and *shape* are also etymologically linked.'[6] Thus, like other '-ships' such as fellowship, comradeship or stewardship, and some archaic terms such as 'countryship' and 'folcship' (which means nation), what is expressed is a set of shared practices and values that are mutually enacted in places, and which contribute to shaping the form of those places.[7]

The relation*ship* between form, communication, values and the performance of belonging are also embedded in landscape (land*ship*). In the research that Furtherfield conducted around the idea of the commons, it became necessary to scaffold between landship, citizenship and the digital realms of 'netizenship' or 'netship'. All these are both separately and mutually constituted, and are increasingly interpenetrating to form an ever-widening ecology and hybrid geography of the situated digital.

Ruth Catlow explains how these concepts, of the commons as landship and netship, are also held in a platform:

> The concept of the online platform is familiar to most people. Through the webpages of social media platforms we 'share' private details of our lives. We add and retrieve records of daily experiences. In recent years we have been taught to mistrust the ways in which our interactions are monitored, our communications tracked, and their value mined for profit or for corrupt political ends. The commons is an invaluable idea because communities steward and organise the resources that they most value, on their own terms.[8]

The commons as both platform and relationship is a vehicle to build solidarity and resist exploitation from without. In a very local way, this helps prevent the slow erosion of access to the things people and other species need in order to prosper. Catlow elaborates:

> We are inviting people to think about the fields of our public park, and the people at play in a similar way. When we visit a park we create new connections and shared memories. Through artworks like Elsa James's *Circle of Blackness* [commissioned for Furtherfield's exhibition 'Time Portals' (2019)] we are inviting people to dig into the histories and lived experiences of local people. The artists of the Transfeminist Rendering Programme will be working this Summer with park users to create 3d scans and build intimate pictures of the outlandish society of creatures that live in, and maintain, the soil of the park.[9]

below and bottom: Gretta Louw's ongoing collaboration with artists from the Warnayaka Art Centre, an indigenous-owned and -operated art centre in the Warlpiri community on the edge of Australia's Tanami Desert, consists of an immersive, on-site multimedia installation centred on the traditional *yujuku* or humpy shelter. A site-specific version was created for Furtherfield.

Rachel Jacobs,
The Prediction Machine,
2017

Anticipating future climate-change events is *The Prediction Machine*'s function, and it is paired with a *Promises Machine* which invites a commitment to action based upon the predictions. These works toured Cambridge, Liverpool, Nottingham, Oxfordshire and Cumbria. Jacobs also created the interactive project *Future Machine* as part of the 'Time Portals' exhibition at Furtherfield Gallery in 2019.

Citizen Sci-Fi

Between 2019 and 2022 Platforming Finsbury Park is being delivered via a three-year programme called Citizen Sci-Fi, the aim of which is to crowdsource visions for a Finsbury Park of the future. Like citizen journalism and citizen science, the emphasis is placed on mobilising a distributed group of, in this case, local communities, to engage in data gathering and sharing activities, building shared imaginaries for alternative realities. Each year has a theme, 2019's being 'Time Portals' and coinciding with the 150th anniversary of the park. In 2020 the theme 'Love Machines' will consider the wellbeing of both people and machines, and in 2021 'Land Oracles' will focus on the key issues of the situated digital and its futurity elucidated here. The first project to kick off the first year of the programme, the *Future Machine*, saw artist Rachel Jacobs working with climate scientists and park users to test different ways of gathering data on the park's climate, and to co-design a machine to relay that data in meaningful ways.

Several women are crucial influences for Citizen Sci-Fi, as they were for Reading the Commons, the work of which helped to develop the programme. Donna J Haraway, who writes of the importance of understanding the epistemic and cultural frames from which we create imagined and possible worlds,[10] is one, and her examining of the concepts of figuring and figuration helped the team to think through the imaginative possibilities of imaginaries (figuring) on the shaping of the world itself (figuration). Great creators of science-fiction imaginaries, especially Ursula K Le Guin and Octavia E Butler, helped point to the transformative power of such worldings. Surely collective imaginaries could generate even more power for real emancipatory change? Especially if these collective imaginaries are generated simultaneously in the entangled lived spaces of netships and landships.

In 2013 Furtherfield Gallery hosted *Seeds Underground Party* by Shu Lea Cheang. That same year the European Union adopted a new seed policy which favoured global agribusinesses through intellectual property, making all seeds subject to strict regulation, and restricting seed exchange by seed farmers and savers. What more fundamental destruction of customary landships could there be than a disruption of the generative impulse to plant and to save for the future?

Shu Lea Cheang invited park users to a seed exchange party where packets of seeds changed hands and went underground in the fields around Finsbury Park and beyond. People came and swapped all kinds of seeds from their gardens. Charlotte Frost, Furtherfield's Director, explains: 'This prefigured our platforming programme by creating a convivial event to which all visitors of the park were invited, and using this as a way to alert people and encourage discussion of the creeping restrictions on important freedoms of exchange and trade.'[11]

Multiple 'others' are held within hybrid geographies and the situated digital commons

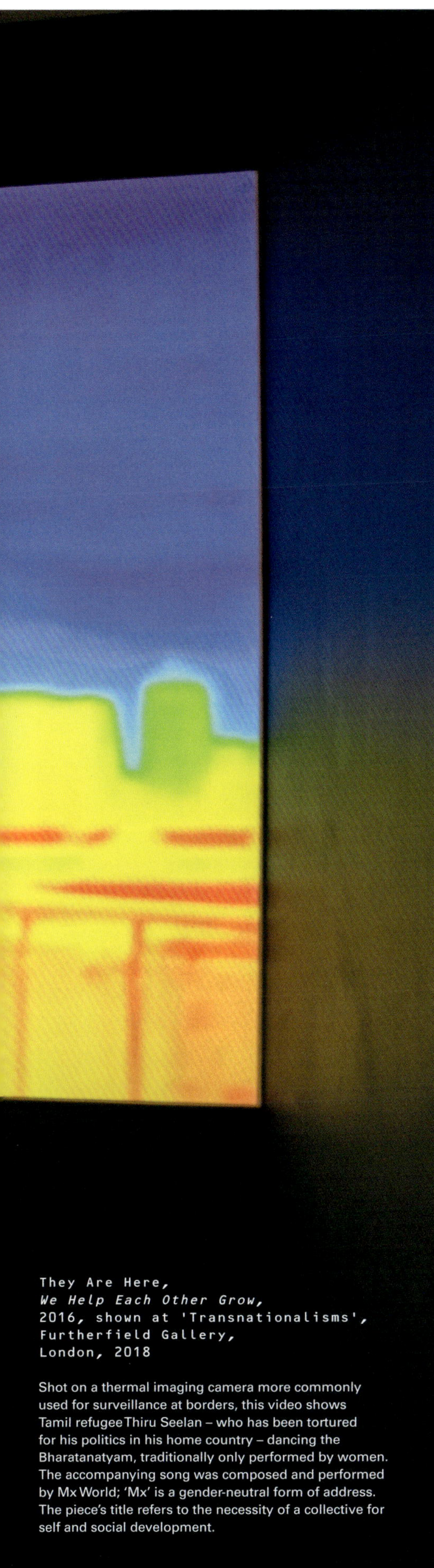

Shot on a thermal imaging camera more commonly used for surveillance at borders, this video shows Tamil refugee Thiru Seelan – who has been tortured for his politics in his home country – dancing the Bharatanatyam, traditionally only performed by women. The accompanying song was composed and performed by Mx World; 'Mx' is a gender-neutral form of address. The piece's title refers to the necessity of a collective for self and social development.

Networking the Unseen in 2016 was the first exhibition of its kind to focus on the intersection of indigenous cultures and zeitgeist digital practices in contemporary art. Curated by artist Gretta Louw and featuring work resulting from many years of collaboration with Neil Jupurrurla Cook and Steve Jampijinpa Patrick, artists of the Warnayaka Arts Centre in Lajamanu, Australia, this show brought together concepts and experiences of remoteness and marginalised cultures, with art making in contemporary society. Once again, it drew attention to the inequalities developing across globalised cultures, and the social and 'cultural impacts of the networks that remain somehow invisible, eroding clearly felt boundaries of geography, place, culture and language'.[12] Further, it drew out importantly that learning from the traditional ecological knowledge (TEK) of indigenous cultures, and how new hybrid geographies have developed from them – these knowledges are far from static – holds significant potential to shape sustainable digital futures.

In 2017 the same programme featured commissions such as They Are Here's *We Help Each Other Grow*. For this four-minute video, a Tamil dance for women, performed by a man, was recorded on a heat-sensitive camera more commonly used in border surveillance to detect smuggled bodies. As though this dancing refugee Thiru Seelan, who learnt the dance in secret from his sister, is only seen for his cultural and geographical transgressions, the work becomes about the policing of public space and traditions. Thiru Seelan's transgression of both gender boundaries and border boundaries is a driver of precisely the sorts of transformations Furtherfield seeks to propose.

Multiple 'others' are held within hybrid geographies and the situated digital commons; the netships and landships within which and through which Furtherfield's work is ever more elaborated. The nonhuman, the more-than-human, the indigenous, the queer, the refugee, are all held within a social ecology that prefigures yet more possibility for the emergence of rich ways of knowing and making that are geared towards planetary flourishing. ⚏

Notes

1. For further discussion of the 'Reading the Commons' project, see Ruth Catlow and Tim Waterman, 'Situating the Digital Commons: A Conversation Between Tim Waterman and Ruth Catlow', *Furtherfield*, 2015: https://www.furtherfield.org/situating-the-digital-commons-a-conversation-between-ruth-catlow-and-tim-waterman/.
2. Owain Jones and Paul Cloke, *Tree Cultures: The Place of Trees and Trees in Their Place*, Berg (Oxford and New York), 2002, *passim*.
3. Sarah Whatmore, 'Heterogeneous Geographies: Reimagining the Spaces of N/nature', in Ian Cook, David Crouch, Simon Naylor and James R Ryan (eds), *Cultural Turns/Geographical Turns: Perspectives on Cultural Geography*, Prentice Hall (Harlow), 2000, pp 265–72, at p 266.
4. *Ibid*, p 267.
5. Ruth Catlow, Marc Garrett, Charlotte Frost and Tim Waterman, live correspondence using Google Docs and email, 2 June 2019.
6. Kenneth R Olwig, *The Meanings of Landscape: Essays on Place, Space, Environment and Justice*, Routledge (London and New York), 2019, p 25.
7. *Ibid*.
8. Ruth Catlow, Marc Garrett, Charlotte Frost and Tim Waterman, live correspondence using Google Docs and email, 2 June 2019.
9. *Ibid*.
10. Donna J Haraway, *Staying with the Trouble: Making Kin in the Chthulucene*, Duke University Press (Durham, NC and London), 2016.
11. Ruth Catlow, Marc Garrett, Charlotte Frost and Tim Waterman, live correspondence using Google Docs and email, 2 June 2019.
12. 'Networking the Unseen: The First Exhibition of its Kind to Focus on the Intersection of Indigenous Cultures and Zeitgeist Digital Practices in Contemporary Art', Art Licks, undated: http://www.artlicks.com/events/5790/networking-the-unseen.

James Corner

Landscape City

Infrastructure, Natural Systems and City-Making

James Corner,
Dry Farming Strips (map collage),
Montana,
1996

Grids, contours, soils and hydrology determine a seemingly
simple yet sophisticated system of strip farming across
massive areas of the windswept Northern Plains of the US.

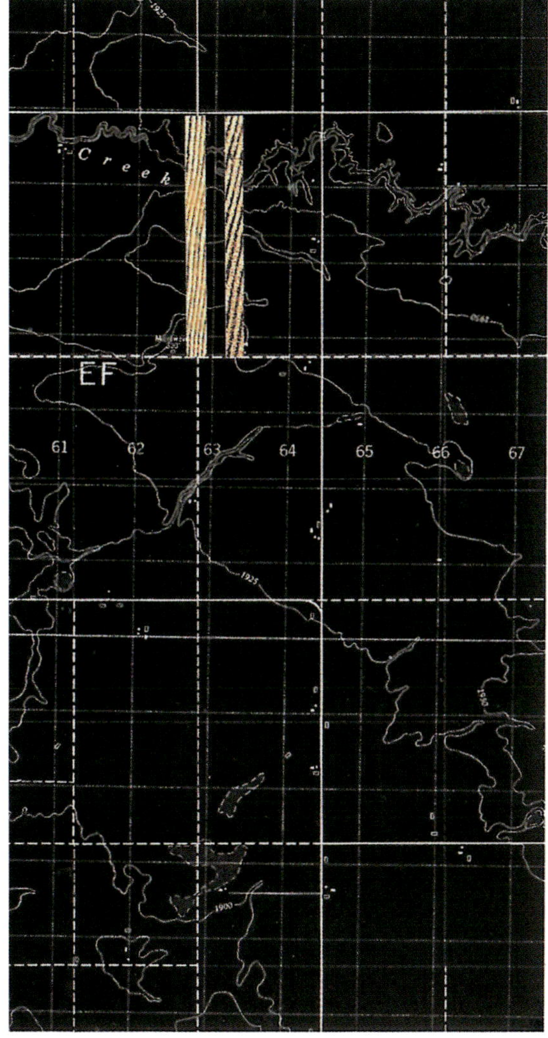

How is current landscape
thinking contributing to
21st-century city design?
Contrasting new ecological
paradigms of resilience with
his older work on the American
landscape of the mid-1990s,
James Corner, founding partner
of Field Operations, explains
the landscape/urban strategy
behind the practice's designs
for China's Xiongan New Area.

The formal and geometrical patterns of the US landscape
are expediently instrumental; they facilitate mobility,
interconnection, settlement, agriculture, industry, extraction,
development and infrastructure. There is little romance or
layering of historical patterns and evolved husbandry, as
might be found in the much older landscapes of Europe and
the UK – just a youthful stridency, directly functional and
straightforward. In 1996, this perspective was the inspiration
for the project Taking Measures Across the American
Landscape, which looked at many different forms, patterns
and geometries inscribed across the land as seen from the
air to try to understand or explain the rationale behind them.[1]
Function, instrumentality, pragmatism and the landscape
at work were foregrounded – a technological landscape of
free-market economics, capitalism, democracy and eclectic
individualism. Low oblique aerial photographs illustrated
the scene, while map collages highlighted the kinds of
geometries, metrics and techniques used to construct and
shape these remarkable landscapes.

Landscape as Infrastructure

Today, nearly two and a half decades later, greater awareness of ecological issues is reshaping the American landscape. Collective concern for climate change, carbon emissions, infrastructure, sustainability and environmental resiliency is leading to new kinds of measures, investment and value in the formation of the landscape. The landscape is urbanising as it gets rewired for the challenges of our time. And, importantly, cities are becoming more like landscapes – not the pastoral landscapes of yesteryear, but the technological landscapes of grids, infrastructures, matrices, webs and working ecologies, all intensely compacted into complex dynamic systems that are being shaped and designed to be more holistic, comprehensive and resilient than ever before.

At first sight, it might not seem particularly convincing to assert that the contemporary city resembles a kind of landscape. Landscape is more popularly thought of as the antithesis of the city, its counterpart, comprised of bucolic countryside and natural areas. Grids, streets, blocks, buildings and freeways are engineered constructions that are quite the opposite of the traditional landscape. There may well be beautiful parks, green waterfronts, squares and gardens that pepper the concrete mass with nature and respite, but the overall urban conglomerate fails to reckon as a singular landscape. Perhaps, at least metaphorically, we might see the city as a kind of geological landscape – stratified, blocky and replete with canyons, pinnacles and promontories – but even this does not seem particularly resonant or useful.

Cities do not 'look' like landscapes, neither in appearance nor material morphology. However, they 'work' like landscapes; their processes, dynamics, interactions and cycles are not unlike landscape ecosystems, layered, temporal and inter-relational. This spatio-temporal perspective underlies the idea of 'landscape urbanism', a way of seeing and acting in the urban field that values time-based interactions and systems over the static space of objects.

Landscape as City

As with ecology and biology, the city can be viewed as a living, working system. The geometrical frameworks that support and facilitate the life of the city also form working landscapes: points, lines, grids, axes, lots, corridors, patches and clusters are as common to landscape as they are to city formation, and these same structures support the various processes of life forms at work.

Louis Kahn once famously likened city streets to water systems, where larger expressways and streets are 'rivers', fed by smaller 'streams', in turn fed by 'canals' and 'docks' with parking structures serving as 'harbours'.[2] These are landscape metaphors speaking more to temporal function than to spatial likeness. Frank Lloyd Wright's unbuilt Broadacre City (1932) painted a mostly green grid mosaic of functioning mixed-use blocks comprising intricate amalgams of

In this conceptual sketch for a new city of over 200 square kilometres (80 square miles), a porous, lattice-like framework maintains the environmental integrity of natural systems and corridors while framing new enclosures of city, town and village, sited according to density and adjacency to transportation systems.

In Field Operations' proposal, the grid is deployed as the simplest means of organisation, setting up a mobility framework as well as development sectors and blocks

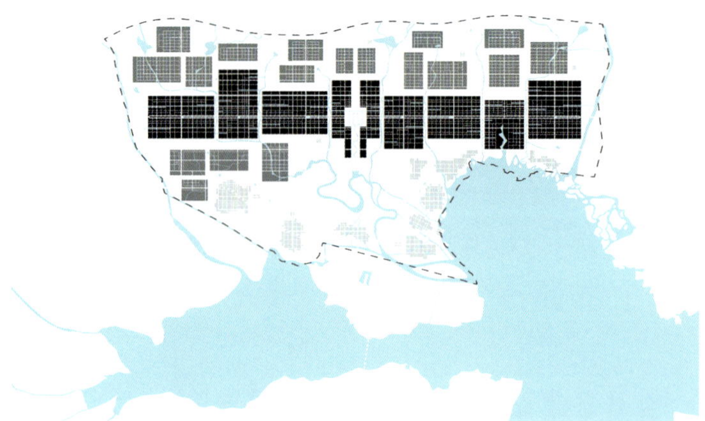

Development plan. The grid allows for maximum connectivity and mobility, while assuming varied dimensions and scales that ensure mixture and texture of building type and programme.

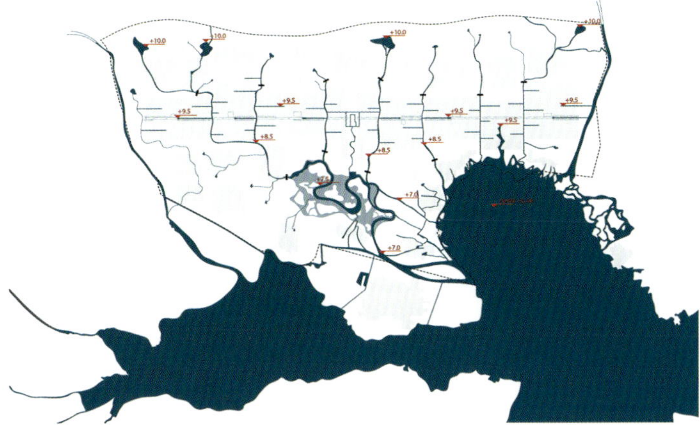

The continuity of the watersheds and water flow systems are coordinated with topography and future development. Water is challenging here because of dry summers and wet monsoon seasons, alternating between drought and flood.

The grid is not continuous or even; instead, it is divided into different areas (or 'towns') with varied scales and different block formats to encourage mixed-use and texture

Open space systems form a massive contiguous fabric, allowing for continuity of ecological systems and flows of matter and energy. The fabric shapes and edges the boundaries of the various city sectors and neighbourhoods. Much of the landscape system can be more or less naturalised and designed with ecological processes in mind; while areas that are embedded inside the city matrix assume a high level of physical design and programming for social function.

buildings, gardens, urban farms, parks, pedestrian ways, vehicular streets, transit corridors and other city components as if a holistic fabric; a landscape of interacting structures – a city designed for work and self-reliance.[3]

So perhaps we can at least differentiate between two kinds of urban landscape: first, the familiar green tissue of parks, waterways, squares, gardens, promenades, pathways and public spaces that are so fundamental to local identity, economic value, social equity, recreational amenity, community life, infrastructural resource, ecological function and environmental resiliency; and second, the urban fabric itself, the matrix-like operating system – the composite framework of networks, grids, blocks, interfaces, edges, corridors, passages and other such structures that facilitate movement, flow, interaction, exchange and work.

This double-sided importance of the 'landscape city' – both as green tissue and operating system – is fundamental to environmental sustainability and resiliency in the face of climate change, growth and urban adaptation. Given the rapid environmental and social changes over the last decades, it is clear that cities need to be able to easily adapt to change; they cannot be viewed, managed or designed as rigid, static forms that are unyielding in the face of change. Instead, they need to be as soft and fluid as they are robust and structured. Resilient systems and communities are better able to recover from stress than fixed ones. Suppleness, pliancy, flexibility and adaptability are key aspects of any resilient system.

Xiongan New Area

The work of Field Operations has long been tied to urban issues. One project that best illustrates the above themes of the 'landscape city' is the practice's proposal for a massive new city in China: Xiongan, just west of Beijing. The Xiongan New Area will eventually become the new administrative centre, relieving pressure on Beijing while also providing much-needed new housing, commercial and institutional facilities. At over 200 square kilometres (80 square miles), such a vast development requires a landscape approach to its overall structuring.

In Field Operations' proposal, the grid is deployed as the simplest means of organisation, setting up a mobility framework as well as development sectors and blocks. However, the grid is not continuous or even; instead, it is divided into different areas (or 'towns') with varied scales and different block formats to encourage mixed-use and texture. This mosaic-like arrangement then allows for the various natural systems to surround and flow continuously from the higher land to the lower central agricultural lands and central lake. This blue-green system enables water to be collected and conveyed during the wet season, and retained and managed during the dry season. Tied to the hydrological scheme are various parks, open spaces and recreational amenities for the new city.

Planned and designed as an operational landscape, the technological frameworks of grids, infrastructures, matrices, webs and working ecologies in Field Operations' proposal for Xiongan are all compacted into intense inter-relational matrices that enable high density combined with open natural systems. The plan integrates green, soft landscapes into the very fabric and infrastructure of the city. This is less a city of objects and more a city of movement, flow, exchange and evolving potential.

The same dialogue between ecological matrix and urban development also scales to each main sector, or district. The key aim is to provide a simple operating system for all urban and environmental processes so that the city can be constructed by myriad developers and agencies over time in a relatively coherent and beneficial manner, with a primary emphasis on a sustainable, resilient and humanistic urban form.

The Xiongan proposal serves as a useful suggestion for how landscape sensibilities can help to inform city building, masterplanning and a more humanistic approach to the design of large-scale multidisciplinary projects. Landscape today continues to embrace countryside and green parks, but has also morphed into the operating system and tissue that undergird urban fabric and its capacity to recover from unforeseen stress and damage. Landscape has become integrally urban, fundamental and crucial to city resilience. Urban planning and design need to better reconcile environmental green systems with urban development, transit mobility and economic models all at the same time. No more silos in terms of professional turf; the city needs to be shaped holistically, conceived and constructed as a total landscape city, an integrated hard and soft weft, an absorptive and facilitative tissue, a porous fabric both yielding and protecting, adapting and evolving in time. The landscape city is a living organism, at once highly efficient, ingenious and beautiful in its transformation and effects.[4] ⏴

Notes

1. James Corner and Alex MacLean, *Taking Measures Across the American Landscape*, Yale University Press (New Haven, CT and London), 1996.
2. Louis Kahn, 'Philadelphia City Planning: Traffic Studies', City of Philadelphia, 1951–3. Drawings and Notes in the Louis I Kahn Collection, Architectural Archives at the University of Pennsylvania, Philadelphia.
3. Frank Lloyd Wright, *The Disappearing City*, WF Payson (New York), 1932.
4. See James Corner, 'Eidetic Operations' and 'Landscape Urbanism', in James Corner and Alison Bick Hirsch (eds), *The Landscape Imagination*, Princeton Architectural Press (New York), 2014, pp 241–56, 291–7.

The landscape city is a living organism, at once highly efficient, ingenious and beautiful in its transformation and effects

WHAT IS DESIGN NOW?

Our era of ecological resilience and ecocide requires much more holistic and inclusive thinking about social, civic space. **Kate Orff** is a co-director of the Center for Resilient Cities at Columbia University Graduate School of Architecture, Planning and Preservation in New York, and founder of landscape and urban design practice SCAPE. Here she outlines the studio's proposal for Alameda Creek, which removes the defensive infrastructures currently restraining the natural forces of the San Francisco Bay Area with the aim of reconnecting its urban and rural ecologies.

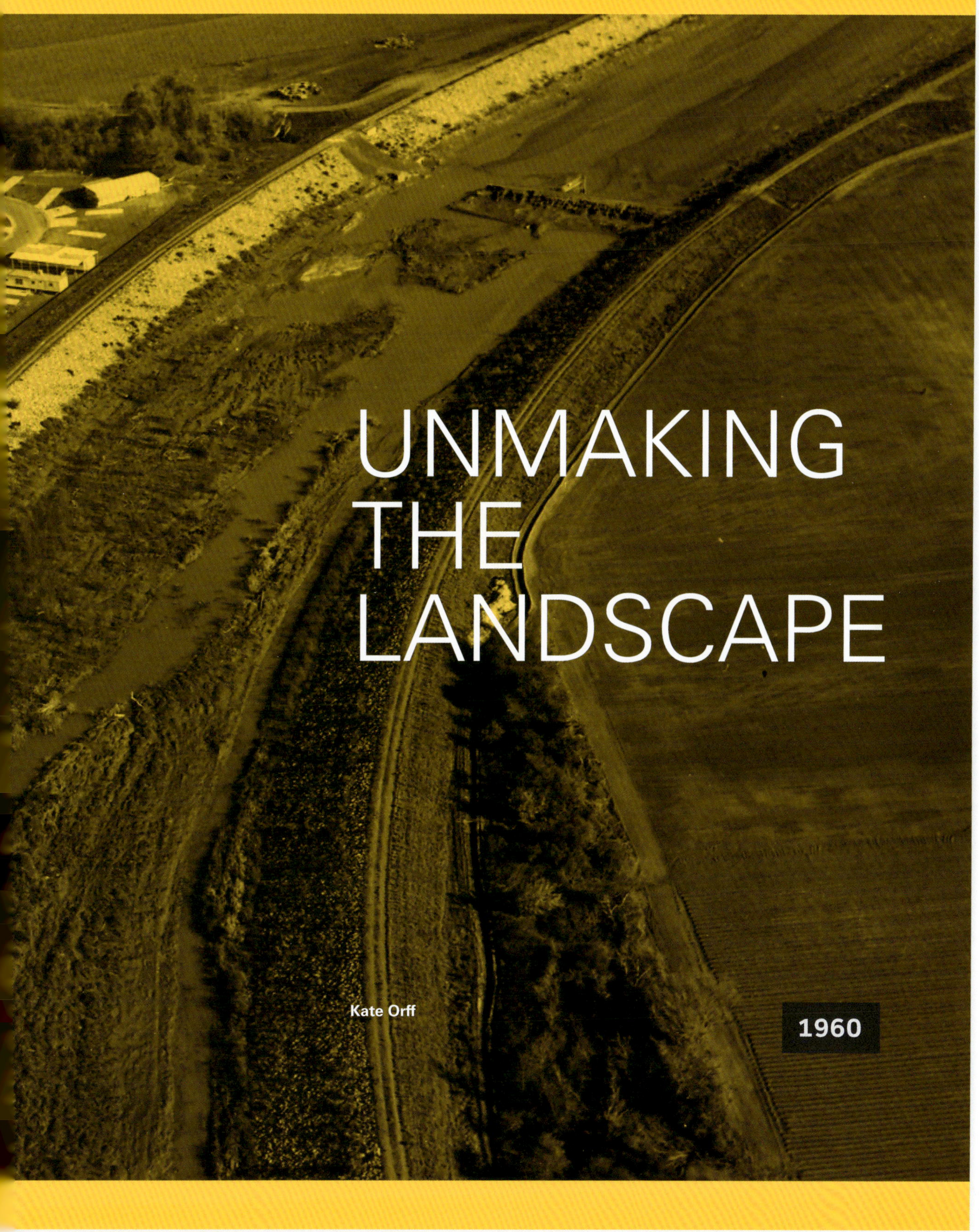

UNMAKING THE LANDSCAPE

Kate Orff

What does it mean to be a landscape architect today, at a moment when the globe is rapidly warming, as the market economy is hitting its zenith, and, at least in the context of the US, civic society is unravelling? Landscape architects are trained to make shapes, to sketch plans, to detail and construct outdoor spaces, but what else do we need to be doing? SCAPE's practice aims to test and expand the agency of the designer.

SCAPE embraces convening, advancing dialogue, and stewardship as deeply creative acts that combine grace and vision, grit and gravitas. Design for the next century is deeply rooted in social life and designing conversations, fostering interaction and encouraging interdependence. Moving forward, in light of increasing climate shocks and stressors, designing the social must be paired with new forms of architectural expression such as un-making, un-doing, subtracting, reversing, decarbonising, tearing out, ripping up, replanting, softening and connecting.

This concept of un-making is evident in SCAPE's Public Sediment for Alameda Creek project in the San Francisco Bay Area of California. The American landscape is dominated by large-scale defensive infrastructure projects that block, channel, divert, dam, harden and channellise rivers and water bodies, from the massive and mighty Mississippi to backyard creeks. What is clear is that these two centuries of 'flood control' and water infrastructure have an unintended legacy of encouraging unsafe development on flood plains, severing fish from spawning grounds, and starving bays of sediment flow. For example, in the nearby Sacramento-San Joaquin delta, the once massively abundant smelt fish is near extinction.[1] In the 2018 species abundance index it was measured as zero. In February 2019, Alameda Creek Alliance volunteers identified exactly one adult steelhead trout in lower Alameda Creek. It was placed in a modest fish tank and driven upstream to its spawning grounds in a van.

In environmental contexts, as with social life, it is seemingly easy to break a fragile system, and it takes a huge amount of regulatory and policy work, convening and collaboration (and driving fish tanks) to reset an integrated dynamic system that rebuilds itself over time. The SCAPE team's Public Sediment project aims to do just that. Like bays and wetlands across the US, the San Francisco Bay Area's tidal ecosystems – its marshes and mudflats – are at risk of subsidence and drowning due to low sediment supply and sea-level rise. Neighbourhoods in the region would lose the protective benefit of adjacent marshlands, and the remaining bathymetry would be inundated until it became flat, open water, leading to a decline in biodiverse intertidal marine life. A large tributary that once fed the Baylands with sediment, Alameda Creek had been transformed from a meander to a shallow and straight concretised canal, cut off from its historical marshes. The project shifts the orientation for sea-level rise adaptation from edge to upstream, redesigning sediment flows to sustain tidal ecosystems and rebuild the protective wetland cushion. The proposal represents a paradigm shift in how we plan for climate change. Rather than hardening the edge, we need to unmake it. This enables a recalibration of our relationship with sediment and water resources, and an investment in living systems that will grow over time to adapt to sea-level rise.

The Public Sediment project was created with extensive input from community stakeholders for the Resilient by Design: Bay Area Challenge initiative launched in September 2017, and has since continued with pilot funding from the National Coastal Resilience Fund. The challenge was a year-long initiative which aimed to address sea-level rise and climate change impacts throughout the area, partially funded by the Rockefeller Foundation and modelled after the Rebuild By Design challenge in New York City. Public Sediment for Alameda Creek is a plan to reconnect the creek with the bay, to provide a sustainable supply of sediment to baylands for sea-level rise adaptation, reconnect migratory fish with their historical spawning grounds, and introduce a network of community spaces that reclaim the Creek as a place for people.

SCAPE,
Public Sediment for Alameda Creek, San Francisco Bay Area, California, 2018

The planning study was built on prior scientific research on subsidence and the loss of tidal wetland ecosystems in the Bay Area, largely due to sea-level rise and sediment supply.

The project proposes the integration of physical and social systems along a revitalised creek to foster more functional and sustainable transportation of sediment, an approach that addresses sea-level rise and encourages public engagement while also providing a habitat for anadromous fish.

The Public Sediment team visualised ongoing scientific research on potential ecological changes in bayland systems over time, focusing on regional-scale losses in sediment supply, fish habitat, and the potential for ecological resilience.

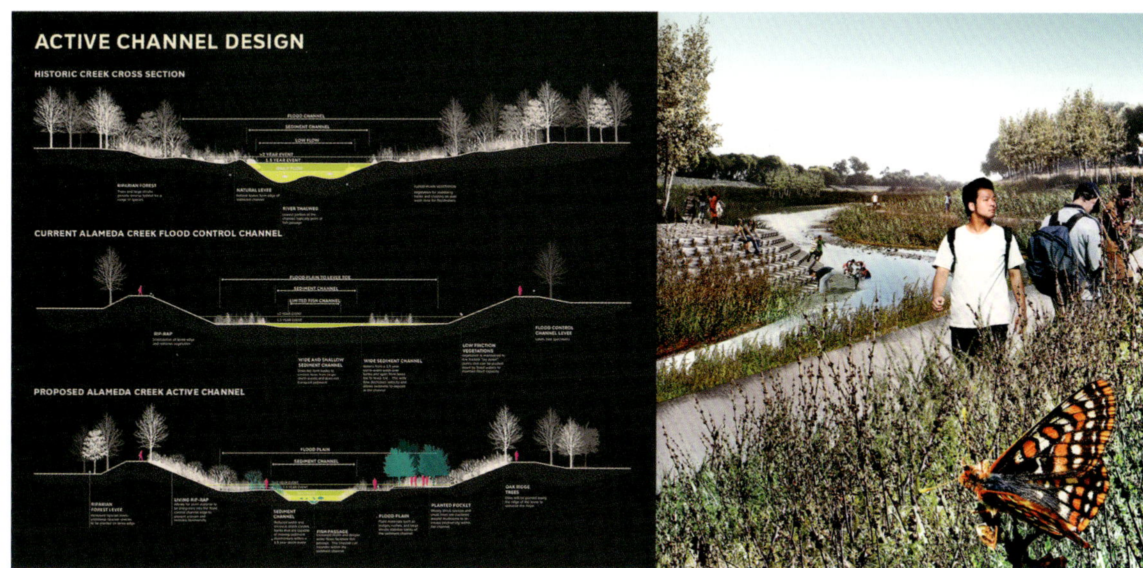

Unmaking years of unsustainable channel design was a main aspect of the Public Sediment proposal. Alameda Creek was once a meandering water body spreading sediment-rich floodwater across a broad floodplain. Following repeat flooding in the mid-1900s, the Creek was channellised for flood protection, leading to ecological decline and disconnection from the water's edge.

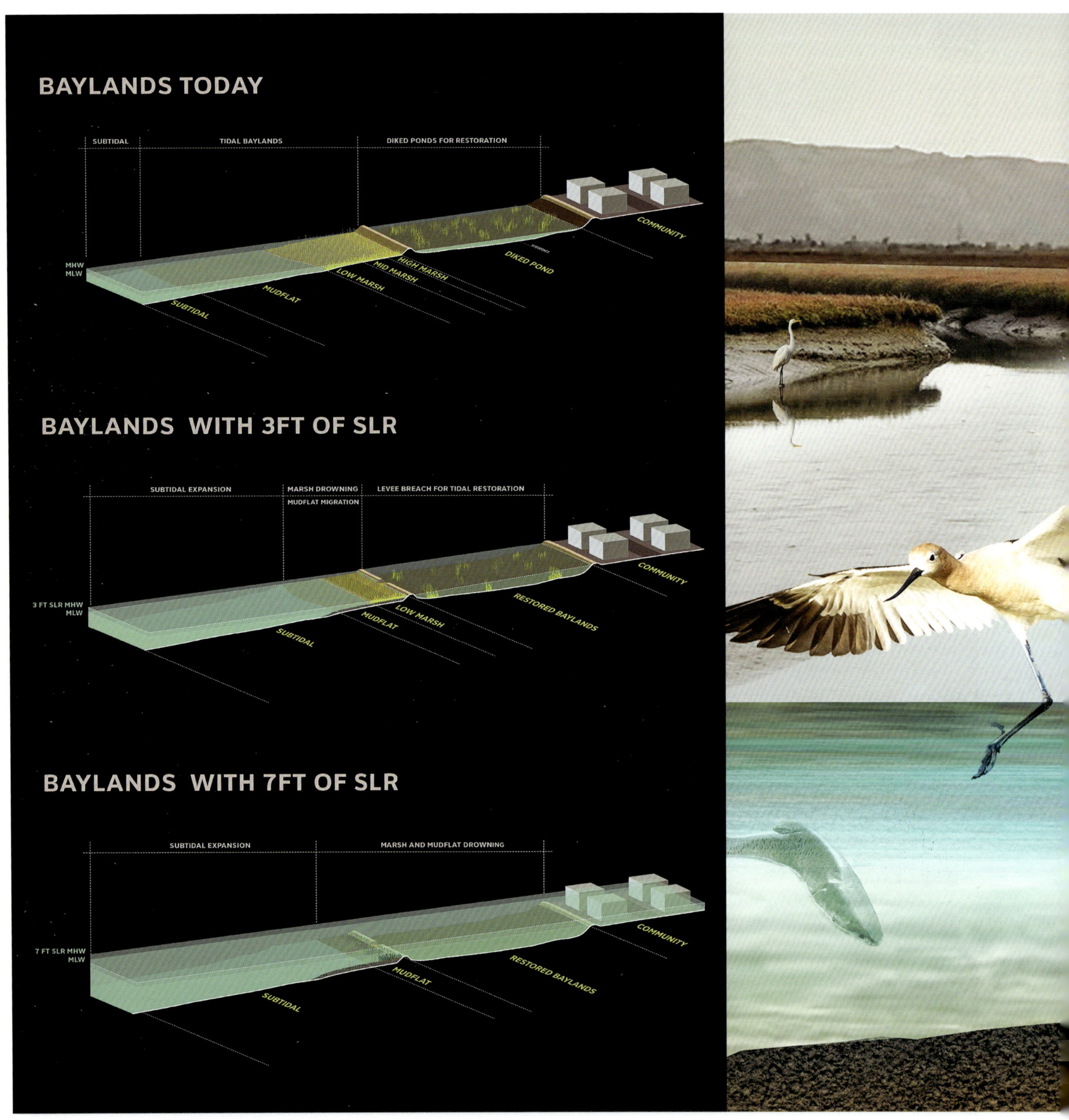

Ecocide and the loss of the earth's biodiversity, in
all its immense texture, colour, scale and intensity,
is an aesthetic and moral question that dwarfs
our more internal conversations about architecture

Adaptive Design in a Global Context

The Center for Resilient Cities and Landscapes at the Columbia University Graduate School of Architecture, Planning and Preservation (GSAPP) has explored these issues in global cities including Amman, Kolkata, Rio de Janeiro and CanTho, among others – places requiring a new, design-driven approach to climate risk. Landscape design now requires new forms of collaboration, and the fostering of political, institutional and creative networks that expand the agency and influence of landscape change. In the Public Sediment project in California, SCAPE teamed up with the Dredge Research Collaborative (DRC), a loose consortium of individuals mostly based in academia in different regions of the US who have studied the larger systems of dredge and the legal and policy landscape, particularly with the US Army Corps of Engineers and state-level regulators. The DRC recognises that these systems have a much larger and more systemic impact on the physical terrain than any site-bounded landscape intervention. SCAPE's goal with the Alameda Creek project, now funded in California's state budget,[2] is to design to influence policy, not just place.

At a global scale, the Center for Resilient Cities and Landscapes marshals the university's design and science expertise to help communities and ecosystems adapt to the pressures of urbanisation, inequality and climate uncertainty. The Center works with public, nonprofit and academic partners to deliver practical and forward-thinking technical assistance that advances project implementation through interdisciplinary research, risk visualisation, project design scenarios and facilitated convenings. It also integrates resilience thinking into design education and academic programming, in particular the Urban Design studio, bringing real-world challenges into the classroom to train future design leaders. These broader academic exchanges can help build joint capacity to design, position and fund transformational initiatives.

This combination of radically unbuilding aspects of the formed landscape, from concretised streambeds to steel bulkheads, levees and seawalls, and collaborating to formulate science, policy and design coalitions to put in place more flexible and adaptable ecological and social methods of addressing climate-changed landscape, are the hallmarks of SCAPE's work. Ecocide and the loss of the earth's biodiversity, in all its immense texture, colour, scale and intensity, is an aesthetic and moral question that dwarfs our more internal conversations about architecture. The act of unmaking the errors of the past, of gathering, and of recognising each other and the earth as worthy of deep care is one of the most profound design challenges before us. ∆

Public Sediment addresses the challenge of sediment scarcity along the vulnerable urban edges of Fremont, Union City and Newark, the vulnerabilities of which will only increase with projected sea-level rise. Beyond the Creek's path, the team's proposal also stretches through tidal zones into the Bay.

Notes
1. See Dan Bacher, 'On Extinction's Edge: Fall Fish Survey Finds Zero Delta Smelt', CounterPunch, 5 February 2019: www.counterpunch.org/2019/02/05/on-extinctions-edge-fall-fish-survey-finds-zero-delta-smelt/.
2. See Senator Bob Wieckowski, 'Wieckowski Highlights Alameda Creek Restoration, Sabercat Trail Funding in State Budget', 28 June 2019: https://sd10.senate.ca.gov/news/2019-06-28-wieckowski-highlights-alameda-creek-restoration-sabercat-trail-funding-state-budget.

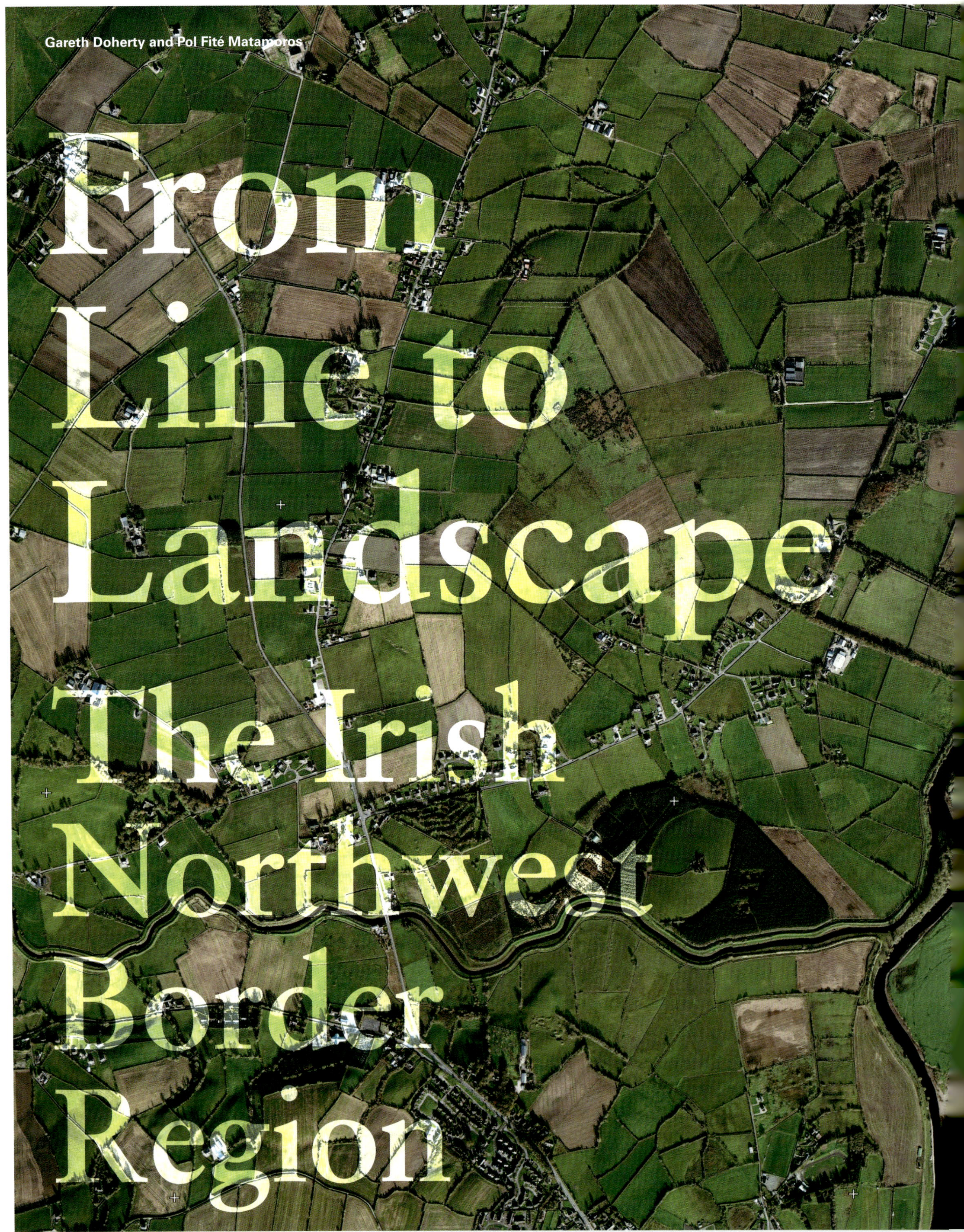
Gareth Doherty and Pol Fité Matamoros

From Line to Landscape

The Irish Northwest Border Region

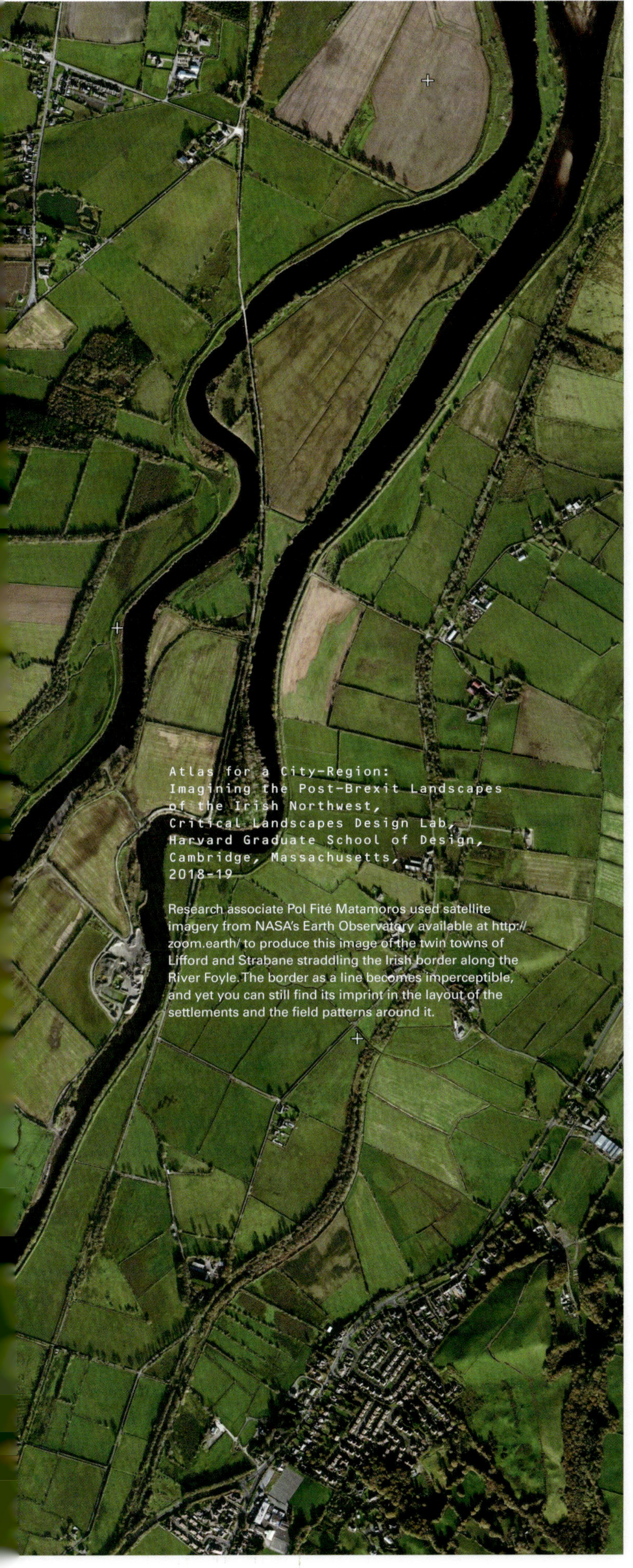

The complexity of border landscapes far exceeds the simple drawing of a line on a map. **Gareth Doherty and Pol Fité Matamoros** discuss the work of the Critical Landscapes Design Lab at Harvard University Graduate School of Design in respect to the Irish/Northern Irish border – a landscape that in recent times has been a fulcrum for important epistemological issues and political machinations.

A border is not a line, it is a landscape. And, to the extent to which borders are landscapes, they mediate – and are produced by – a myriad of interrelationships that span space, scale and time, humans, animals and plant life, blurring the precision of political lines and complicating the binaries that too often define them. Taking a 'landscape approach' to border territories is a productive conceptual and methodological manoeuvre to critically interrogate and design in such contested, politicised and increasingly topical contexts. But what does it mean to take a 'landscape approach' to a border context, and specifically the Ireland/Northern Ireland border which has been such an integral part of the British political landscape from partition in 1921–2, through to the discussions over the UK's withdrawal from the European Union?

The Irish Northwest as a Border Landscape
Over the course of 2018–19, the Critical Landscapes Design Lab at the Harvard Graduate School of Design engaged in a landscape approach to the Irish/Northern Irish border territory through a sponsored research project entitled 'Atlas for a City-Region: Imagining the Post-Brexit Landscapes of the Irish Northwest'. The approach enabled the Design Lab to look beyond the border itself – the manifestation of an intractable political situation – and focus on the systems of interrelationships that occur around and across the border, as well as those that supersede it. It forces one to challenge the monolithism and purity of the political line drawn on a map to recognise that a border is, in fact, made of much more than a line. How can we not consider the fluidity of a border when, in the Irish Northwest, it runs along the River Foyle for about 10 miles (16 kilometres) and regularly floods? Political borders do not easily resist streams, storms, bees or transnational corporations, never mind the everyday lives of inhabitants on either side.

However, a border is not just a membrane of diverse porosities. As with any other landscape element, it establishes a dialectical relationship with its environment, both shaping and being shaped by it. Hence, if we recognise the imprint of the border in the built and unbuilt environment in the same way we do that of a river, the border as a discrete line needs to be reconsidered as a much more diffuse and staggered entity. It needs to thicken. It needs to become a landscape.

Thus, even in the context of the soft – almost non-perceptible – border between Ireland and Northern Ireland, the border condition of its landscapes continues to be materialised and spatialised through its two different housing markets, its two asymmetrical infrastructural plans and health services, and even the spread of petrol stations, lotto agents and supermarkets that benefit from operating with two different currencies.

This border landscape is, in fact, a lived landscape. And a landscape approach that focuses on the systems of interrelationships that comprise it cannot – and must not – disentangle the human from the nonhuman.[1] More accurately, as Félix Guattari writes in *The Three Ecologies* (1989), the environmental should not be considered in isolation from the social and subjective ecologies.[2]

In this context, the thickening of the border from line to landscape can also help to break the political and social binaries through which border territories are too often simplified and addressed. How can both sides of a border be considered as discrete and opposite entities when the line that divides them has become a thick and diffuse landscape? The dialectical relationship between border and environment must then also apply to its people, generating a distinct set of border ecologies and economies that allow us to address border landscapes as a third and non-binary category of its own.

Thus, the Irish border in the Northwest not only manifests through holiday homes, petrol stations and electricity networks. It also shapes how people live their everyday lives on a cross-border basis – from grocery shopping, to going to school, to filling the tank of the car – how they organise their communities, and how they develop something similar to a regional identity. The border, then, also needs to be considered as a collective imaginary that not only produces distinct and shared identification marks – such as city murals, music or tales – but also enables different border communities to externalise their political and religious differences: there cannot be a border *here* if it is *there*.

Killea is a commuting town that sits on the Irish border and has grown in recent years because of its strategic location. This photograph, taken by Master in Urban Planning student McKayla Dunfey, depicts the border store and gas station there – one of many that punctuate these border towns.

The Irish border along the River Foyle and its facades, Counties Donegal and Tyrone, photographed by Master in Architecture student Sohun Kang. The border runs along the river for about 10 miles (16 kilometres) in the Irish Northwest and between the twin towns of Lifford and Strabane, which suffered severe flooding over the years.

A hedgerow in Inishowen, County Donegal, photographed by Master in Landscape Architecture student Joan Chen. Hedgerows are one of the most pervasive elements in this region's landscape. They are both borders that delimit private property and ecological corridors. They are both membranes of diverse porosities and a fundamental element of these landscapes that establish a dialectical relationship with its environment.

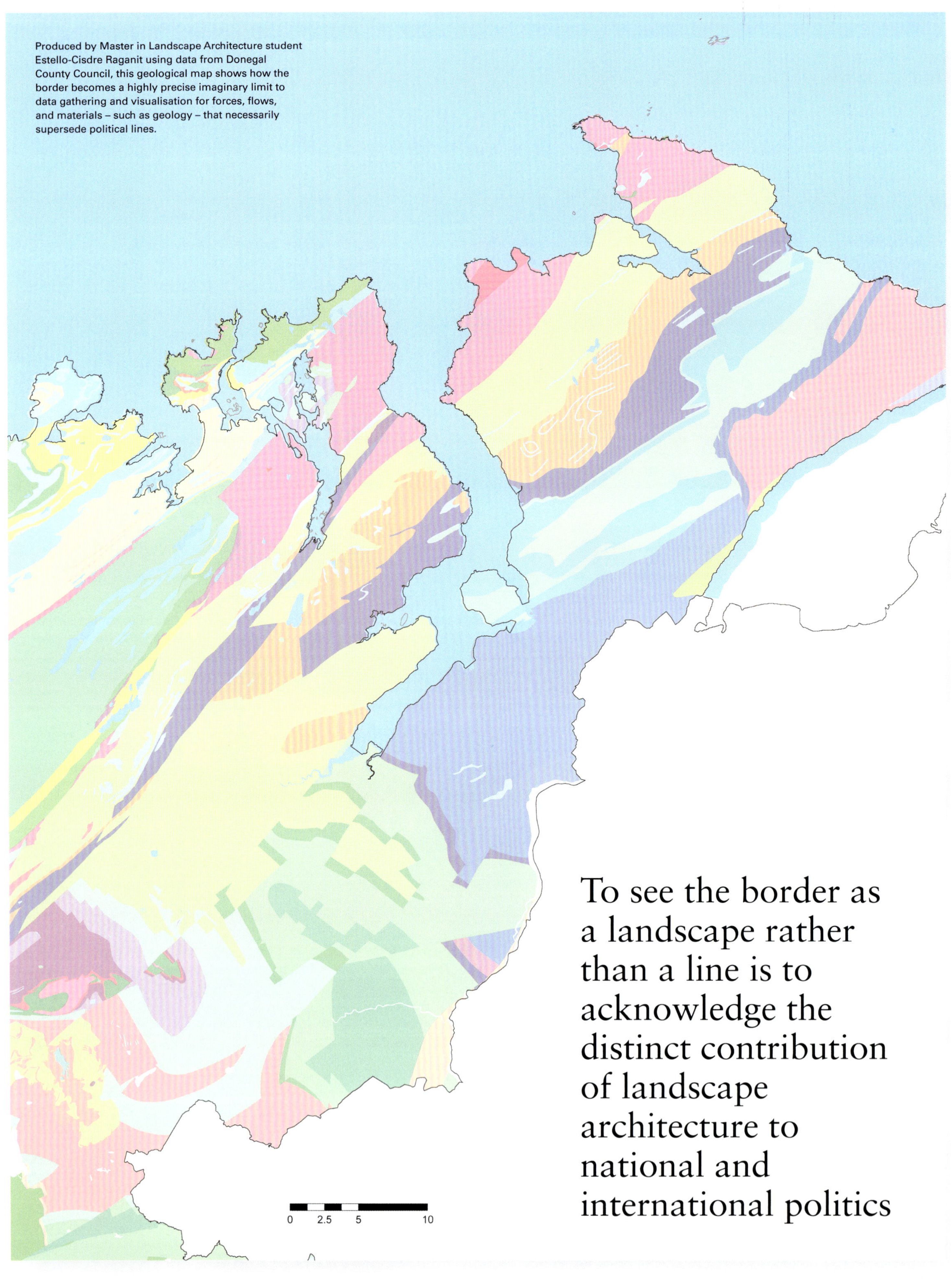

To see the border as a landscape rather than a line is to acknowledge the distinct contribution of landscape architecture to national and international politics

Engaging in Landscape Fieldwork

In order to mobilise this landscape approach, fieldwork based on ethnographic interactions is a fundamental method that can elucidate such interrelationships, especially in a border context. Mainstream understandings of borders rely too heavily on census data, derived through different sets of criteria on either side of a line. The limits to using such forms of census data and their predetermined categories of analysis are only enhanced when dealing with two different jurisdictions – and their different legibility projects – that, in their imaginaries, only share a line: the border.[3] This complexity of interrelationships is then reduced to a mere quantification of certain forms of trade and immigration, crystallised in an abstract polygonal line.

How can we even begin to think about cross-border mobility when census data only records travel-to-work time and disregards other forms of everyday cross-border mobility? How can we talk about food resilience if we do not understand the interdependences between Irish and Northern Irish farmers? How can we design with border towns and villages if we ignore how they have shaped their collective identities and community ties? Touching, smelling, walking, talking, recording, drawing – all these actions become fundamental steps towards generating forms of data that can provide much more grounded and nuanced understandings of how borders operate as landscapes, and how to design with and for these landscapes.

Over the course of one week, 17 Design Anthropology students from Harvard University descended on the Irish Northwest to conduct fieldwork, living in communities and farms and engaging in everyday life as far as possible throughout the border landscape in both political jurisdictions. The goal was to understand not only the spatial and material characteristics that constitute the Irish Northwest – and its border – but also the set of forces, flows, dreams, fears and hopes that shape them and do not normally show up in official statistics. The border, then, becomes a much more complex and multifaceted reality that operates as both a divider and an enabler, a boundary as well as an opening, extending its imprint much beyond the political line and its checkpoints.[4] The border becomes a way of living that extends to towns, villages, farms, fields and rivers, even to those that appear to be removed from it. The border becomes a landscape.

To see the border as a landscape rather than a line is to acknowledge the distinct contribution of landscape architecture to national and international politics. It is a recognition of the liminality of lines but also the multiple borders that exist within and adjacent to the line, including those that are inside our minds – which, as one of the project's interlocutors said, 'is the most important border of all'. ⌂

Notes
1. For more on landscape definitions, see Anne Whiston Spirn, *The Language of Landscape*, Yale University Press (New Haven, CT), 1998, pp 24–5.
2. Félix Guattari, *The Three Ecologies* [1989], trans Ian Pindar and Paul Sutton, Athlone Press (London and New Brunswick, NJ), 2000, p 28.
3. James C Scott, 'State Simplifications: Nature, Space and People', *Journal of Political Philosophy*, 3 (3), 1995, pp 191–233.
4. For more on borders and boundaries, see Richard Sennett, 'Edges: Self and City', in Mohsen Mostafavi (ed), *Ethics of the Urban: The City and the Spaces of the Political*, Lars Müller Publishers (Zurich), 2017, pp 261–8.

View from Shore Walk bench across Lough Foyle, in real life and in a watercolour sketch by Master in Landscape Architecture student Joan Chen. Ethnographic fieldwork prevents us from being too detached from the landscape we are trying to design with. Walking, touching, and drawing then become key-methods to overcome the limits to normative census data and produce other understandings of the landscape.

Matthew Gandy

AT

TANG

DELINEATING A NEW ECOLOGICAL IMAGINARY

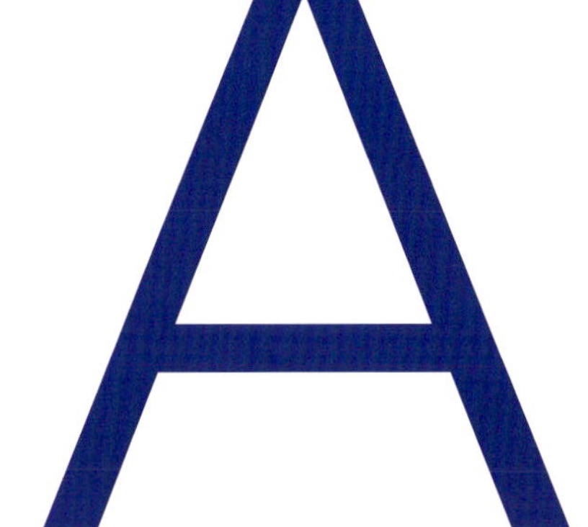

A

ENT

How are the 'iron landscapes'
of disused railway tracks of
the 19th and 20th centuries
rejuvenated by encouraging
inclusive and diverse uses?
Matthew Gandy, a professor
of geography at the University
of Cambridge, charts the
phased design of the Park
am Gleisdreieck in Berlin. He
questions whether the design
of the park can be considered
separately from the politics
of housing as he highlights
tensions between civic society
and urban regeneration.

A somewhat bleak and nondescript area of disused railway lines, marking a boundary between the former West Berlin districts of Kreuzberg and Schoeneberg, has recently become the focus of international attention through the creation of a new public park. Park am Gleisdreieck, the first section of which was completed in 2011, is now filled with people on warm afternoons. The park design is notable for its sensitivity to the independent ecological dynamics of urban nature as well as a degree of inclusivity towards subcultural dimensions to Berlin life such as street art. Yet even the most sophisticated designs necessarily emerge within a specific set of constraints, and in the case of Berlin this wider context includes the steady eradication of marginal spaces of cultural and ecological significance as well as the rapidly rising costs of housing.

The complex negotiations that have enabled this new public space to be created have occurred within a context of increasing financialisation within the housing sector across Berlin and other German cities.[1] The question of park design cannot be considered separately from the politics of housing provision and the impact of rising levels of socioeconomic inequality. An existing form of 'social contract' between civil society and a variety of housing providers, both in the private and public sectors, has become severely frayed.

View of a newly completed section of the park with traces of industrial archaeology and ruderal vegetation.

Track Wilderness

The name Gleisdreieck, meaning 'track triangle', actually sounds strange in German as well as English. The German writer and actor Hanns Zischler notes how Gleisdreieck is a 'strange word, the name of a location without a location, like a technical-geometrical paradox'.[2] The site originates from a complex track formation created in 1912 to improve the flow of trains through the rapidly expanding city of Berlin. Its brutal functionality was captured in the Austrian writer Joseph Roth's expression 'iron landscape', from an essay published in 1924, where he referred to the space as a 'playground of machines'.[3] The array of tracks also features in Walter Ruttmann's classic evocation of Weimar modernity in his film *Berlin: Sinfonie der Großstadt* (*Berlin: Symphony of a Metropolis*) (1927) and in Robert Stemmle's Nazi-era film *Gleisdreieck* (1936). In 1942 the railway lines became part of the infrastructure of annihilation that deported Jewish people and other minorities to their deaths. Towards the close of the Second World War the system of tracks and stations was heavily damaged by aerial bombardment. By the early 1950s the site had been decommissioned as a transport hub, along with a series of other connecting spaces, and Gleisdreieck became part of the patchwork of anomalous and abandoned spaces that came to characterise postwar Berlin. The track complex, with its assortment of ruined buildings, became a prominent component in the truncated and marooned infrastructure networks that characterised the divided city.[4] Over time the site began to acquire significance as a kind of 'vernacular park', imprinted into grassroots forms of collective memory. Gleisdreieck and other ruined spaces became a playground for social, cultural and sexual experimentation.[5]

Located in the former island city of West Berlin, the sprawling Gleisdreieck site became a zone of contestation in the early 1970s as a citizens' initiative was established to oppose the building of a new motorway through the site under the so-called Westtangente ('West Tangent') scheme. This ultimately unrealised proposal reflects the postwar emphasis on the *autogerechte Stadt* ('the car-corrected city') as a leitmotif for urban planning. By the late 1970s local opposition to the road-building scheme had developed into a sophisticated alternative plan for a linear park through the inner city called the Grüntangente ('Green Tangent') that would have connected all the way from Gleisdreieck to what is now the speculative commercial hub of Potsdamerplatz, as well as incorporating abandoned railway lands to the south.

Once a symbol of industrial modernity, the now heavily overgrown landscapes of Gleisdreieck became emblematic of the *terrain vague* of postwar Berlin. In films such as Wim Wenders's *Der Himmel über Berlin* (*Wings of Desire*) (1987) the ostensibly empty spaces of Gleisdreieck signal a fracturing of memory within a city cut adrift by geopolitical separation. Yet Wenders's evocation of a time and space out of sync must be read alongside a parallel fascination with the site as a novel kind of socioecological assemblage. A now classic botanical study from 1980 by Ulrich Asmus recorded over 400 different species of plants across the site as part of a newly articulated emphasis on urban biodiversity led by Herbert Sukopp, Ingo Kowarik and other ecologists based at Berlin's Technical University.[6] For urban botanists, sites such as Gleisdreieck exemplified a kind of *Stadtwildnis* (urban wilderness) marked by unusual combinations of native and non-native species. Implicitly, therefore, this recognition of new kinds of 'cosmopolitan ecologies' marked a point of departure from existing approaches to the study of plant succession and identifiable vegetation types.

During the 1980s, this scientific lens emanating from the Technical University was to acquire increasing political significance under what the sociologist Jens Lachmund terms an emerging 'biotope-protection regime' that became a distinctive feature of environmental politics and land-use planning in West Berlin.[7] The eventual creation of a new public park on the Gleisdreieck site marks a kind of cultural, political and scientific continuity with strands of social and environmental activism in the former West Berlin, mirroring other recent projects such as the Südgelände urban nature park, and underpinned by the pedagogic connections between former students on the urban ecology programme at the Technical University and the design studio Atelier LOIDL that won the park design competition held in 2006 after seeing off 85 other design proposals.

Atelier LOIDL,
Park am Gleisdreieck,
Berlin,
2019

The park is traversed by sections of abandoned railway tracks including strips of urban woodland referred to as *Gleiswildnis* (track wilderness).

The eventual creation of a new public park on the Gleisdreieck site marks a kind of cultural, political and scientific continuity with strands of social and environmental activism

Functional train lines have been incorporated into the park design.

Urban Ecology as Design Motif

Under construction since 2006, and opening in stages since 2011, the prize-winning design consists of a series of zones or 'rooms' reflecting a diversity of different social, cultural and ecological interests. When approaching the site from the Hornstrasse entrance, on the east side of the park, visitors are welcomed by a series of wide steps, along with a ramp that clearly signals a welcoming and well-designed space. At the top of the steps there is a network of paths into an open space called the Kreuzberger Wiese (Kreuzberg meadow). Interestingly, this space is divided between lawn-like areas that become extremely popular on warm days, combined with more lush strips of vegetation enabled by differential mowing regimes. These demarcated patterns are reminiscent of Gilles Clément's design for Parc Henri Matisse (1995) in Lille, and serve a useful role in demonstrating how different vegetation types are intentional rather than a byproduct of neglect or mismanagement: the use of sharply edged contrasts allows different types of spontaneous nature to be aesthetically staged as part of the larger design concept for the park.[8]

Similarly, the park makes extensive use of rubble substrates to foster flower-rich ruderal ecologies, including many classic examples of wasteland or *Brache* flora such as *Dipsacus fullonum*, *Echium vulgare* and *Verbascum densiflorum*. These specially provided substrates are clearly labelled as *Ökoschotter* (eco-rubble) to provide a comparable didactic effect to the kind of ecological zones that one might encounter in a botanical garden. Additionally, there are signs for *Gleiswildnis* (track wilderness) that indicate remaining fragments of the spontaneous urban woodland that developed on the site before its conversion into a park. Dominated by birch trees with gnarled trunks that twist their way around the rusty tracks, these post-industrial woodlands provide a material continuity with the characteristics of the space before its formal incorporation into the Berlin park system.

Dotted throughout the park are a variety of sports and recreational facilities including more niche features such as concrete bowls for skateboarding and an officially sanctioned 'graffiti wall'. These highly popular elements of the park design are indicative of a degree of inclusivity for pre-existing urban subcultures that can be read in parallel to the aesthetic appropriation of the ecological characteristics of the original site. Towards the western perimeter there are further features of interest including surviving allotment colonies that predate the creation of the park and have now been encircled within its protective configuration. There are also 'wilder' zones adjacent to a large sign that illustrates a range of interesting flora and fauna including nine different *Stationen des Rundgangs* (tour points) that encourage visitors to experience urban nature as a series of dynamic zones of ecological transition: progressing further into this wilder part of the park there is a network of didactic infrastructure for these different forms of *Stadtwildnis* (urban wilderness) such as *Birken-Pappel-Vorwald* (birch-poplar early-stage forest) and *Pioneer Gebüsche (mit Sanddorn)* (pioneer shrubs with sea buckthorn). It is striking though how deserted these western areas of Gleisdreieck feel even on a warm day when the rest of the park is buzzing

In the foreground are species such as *Echium vulgare*, *Oenothera biennis* and *Saponaria officinalis* that are growing in the eco-rubble.

with activity; and deeper within the woods there are traces of homelessness along with the presence of other marginalised people who are seeking shelter from the increasingly expensive and inhospitable city beyond.

Ecological Atmospherics

In terms of its usage, the park is clearly very successful and seems to attract a predominantly young crowd of visitors. Proponents for the park such as the cultural geographer Jürgen Hasse describe its success as 'a space of affective dynamism' marked by diversity of 'atmospheres' and 'situations'. Hasse draws on elements of the so-called 'new phenomenology', and especially the philosophy of Hermann Schmitz, to highlight human interactions with nonhuman elements such as weather or seasons to produce a skein of micro choreographies.[9] Public spaces such as parks have long served as a focal point for ethnographic observations of everyday life, and in particular the presence of what the anthropologist Kathleen Stewart refers to as 'situations' or other kinds of micro-disturbances that ripple through spatial settings. There is now growing interest in 'affective atmospheres' as an alternative vantage point to more narrowly defined conceptions of the human subject encountered in classic accounts of architectural history and design theory.[10]

From an environmental perspective the park provides a rich setting for wider reflections on the role of nature in urban design. The ecologist Ingo Kowarik notes how the park successfully combines what he terms 'third nature', encompassing classic elements of garden design, with 'fourth nature' exemplified by traces of *Stadtwildnis* (urban wilderness).[11] For Kowarik, this synthesis rests principally on the material configuration of the new park design, but we might widen this field of interpretation to consider whether the park signals a new or distinctive kind of ecological imaginary. The deployment of the term 'ecological imaginary' moves beyond material typologies of human interaction with nature to emphasise cultural projections onto the nonhuman realm. In the case of Park am Gleisdreieck the design incorporates radically different elements of urban nature so that any putative ecological imaginary is best conceived as an experimental synthesis. The didactic features of the park resemble an urban botanical garden, while the more human-oriented recreational features, including the network of paths, provide a kind of late-modern promenade that intersects with the industrial archaeology of the site.

In the case of Park am Gleisdreieck the design incorporates radically different elements of urban nature so that any putative ecological imaginary is best conceived as an experimental synthesis

The design includes popular concrete bowls for skateboarding and other recreational activities.

In addition to a designated graffiti wall, many other surfaces are regularly festooned with various types of street art.

The eastern side of the park includes areas of *Stadtwildnis* (urban wilderness) that have been largely left alone and include a series of didactic signs to show different aspects of vegetation dynamics. This sign indicates *Birken-Pappel-Vorwald* (birch-poplar early-stage forest) so that parts of the park resemble an urban botanical garden.

Given the scale and sophistication of the public works required for the construction of the park, it is interesting to consider how such a large and complex kind of public space came into being. The funding for Park am Gleisdreieck has emerged out of negotiations for the destruction of open space elsewhere in the city, notably in Potsdamerplatz, and is a prominent example of what can be termed a 'compensation landscape' that has emerged out of discussions between the state, private interests and community organisations.[12] Furthermore, the management of the park, which at times appears highly interventionist, emerges from a hybrid form of public-private partnership through the operation of a state-owned set of companies operating as Grün Berlin GmbH since 1992. The powerful Grün Berlin entity now oversees key elements of design, project management, maintenance and, increasingly, security for a network of over 1,000 hectares (2,470 acres) of public spaces across the city.

Although different stages of the Gleisdreieck project have sought the involvement of civic society, and used various forms of public consultation, there have been significant tensions between the managerial impetus of Grün Berlin and opposition to the excessive use of asphalt surfaces or other types 'non-ecological' interventions.[13] Furthermore, the security presence appears quite excessive in comparison with other public spaces in Berlin, with regular patrols observable in almost all parts of the park. Above all, the steady advance of high-end housing developments around the perimeter of the park betokens a close elision between urban design and speculative urban development: the sense of a tranquil oasis surrounded by various fragments of *terrain vague* and inexpensive apartments is being steadily displaced by the emergence of an urban canyon not unlike the 20th-century enclosure of Manhattan's Central Park or the more recent gentrification frenzy associated with the High Line.

Park am Gleisdreieck is a paradoxical public space: elements of its design display a high degree of cultural and scientific sophistication, yet its very success has contributed to rapidly rising rents in surrounding neighbourhoods. Since the vast majority of local residents do not own their apartments, any increase in rent may ultimately force many current users of the park to relocate to cheaper parts of the city. These shifts are significant not only in class terms, but also in relation to the ethnic composition of park users, since these areas of former West Berlin had a high proportion of Turkish households as well as other migrant communities. It is instructive, therefore, that a park design that is so sensitive to the cosmopolitan and subcultural characteristics of inner-city Berlin life should nonetheless find itself tangled in the wider dynamics of speculative urban change. The park may succeed in holding relic elements of the original ecological characteristics of the site in a state of suspended animation, but the protection of the social and cultural complexity of such marginal spaces is far less certain. ᴆ

An example of newly constructed luxury housing developments around the edge of the park.

Notes
1. See, for example, Knut Unger, 'Financialization of Rental Mass Housing in Germany: Understanding the Transaction Cycles in the Mass Rental Housing Sector 1999–2015', in Barbara Schönig and Sebastian Schipper (eds), *Urban Austerity: Impacts of the Global Financial Crisis on Cities in Europe*, Theater der Zeit (Berlin), 2016, pp 176–90.
2. Interview with Hanns Zischler in Andra Lichtenstein and Flavia Alice Mameli (eds), *Gleisdreieck/Park Life Berlin*, transcript (Bielefeld), 2015, p 83. See also Hanns Zischler, *Berlin ist zu groß für Berlin*, Galiani (Berlin), 2013.
3. Joseph Roth, 'Bekenntnis zum Gleisdreieck', *Frankfurter Zeitung*, 16 July 1924, in Joseph Roth, *Orte*, Reclam (Leipzig), 1990, p 68.
4. See, for example, Sandra Jasper, 'Phantom Limbs', in Matthew Gandy (ed), *Urban Constellations*, Jovis (Berlin), 2011, pp 153–6.
5. See, for example, Wolfgang Schivelbusch, *In a Cold Crater: Cultural and Intellectual Life in Berlin, 1945–1948*, trans Kelly Barry, University of California Press (Berkeley, CA), 1998.
6. Ulrich Asmus, *Vegetationskundliches Gutachten über den Potsdamer und Anhalter Güterbahnhof in Berlin*, Berlin Senate, 1980.
7. Jens Lachmund, *Greening Berlin: the Co-production of Science, Politics, and Urban Nature*, MIT Press (Cambridge, MA), 2013, p 127.
8. See Matthew Gandy, 'Entropy by Design: Gilles Clément, Parc Henri Matisse and the Limits to Avant-Garde Urbanism', *International Journal of Urban and Regional Research*, 37 (1), 2013, pp 259–78.
9. Jürgen Hasse, 'Zur Atmosphäre eines urbanen Grünraums: Der Park am Gleisdreieck', in Lichtenstein and Mameli, *op cit*, pp 238–43.
10. See Matthew Gandy, 'Urban Atmospheres', *Cultural Geographies*, 24 (3), 2017, pp 353–74.
11. Ingo Kowarik, 'Gleisdreieck: wie urbane Wildnis in neuen Park möglich wurde', in Lichtenstein and Mameli, *op cit*, pp 201–15.
12. Kowarik discusses the creation of new landscapes as a form of compensation in the documentary film *Natura Urbana: the Brachen of Berlin*, director: Matthew Gandy, 2017.
13. Thomas Loy, 'Hauptstadt-Gärtner-das Millionen-Unternehmen "Grün Berlin"', *Der Tagesspiegel*, 26 January 2018: www.tagesspiegel.de/berlin/von-marzahn-bis-neukoelln-hauptstadt-gaertner-das-millionen-unternehmen-gruen-berlin/20863736.html.

Teddy Cruz and
Fonna Forman

NATION AGAINST NATURE

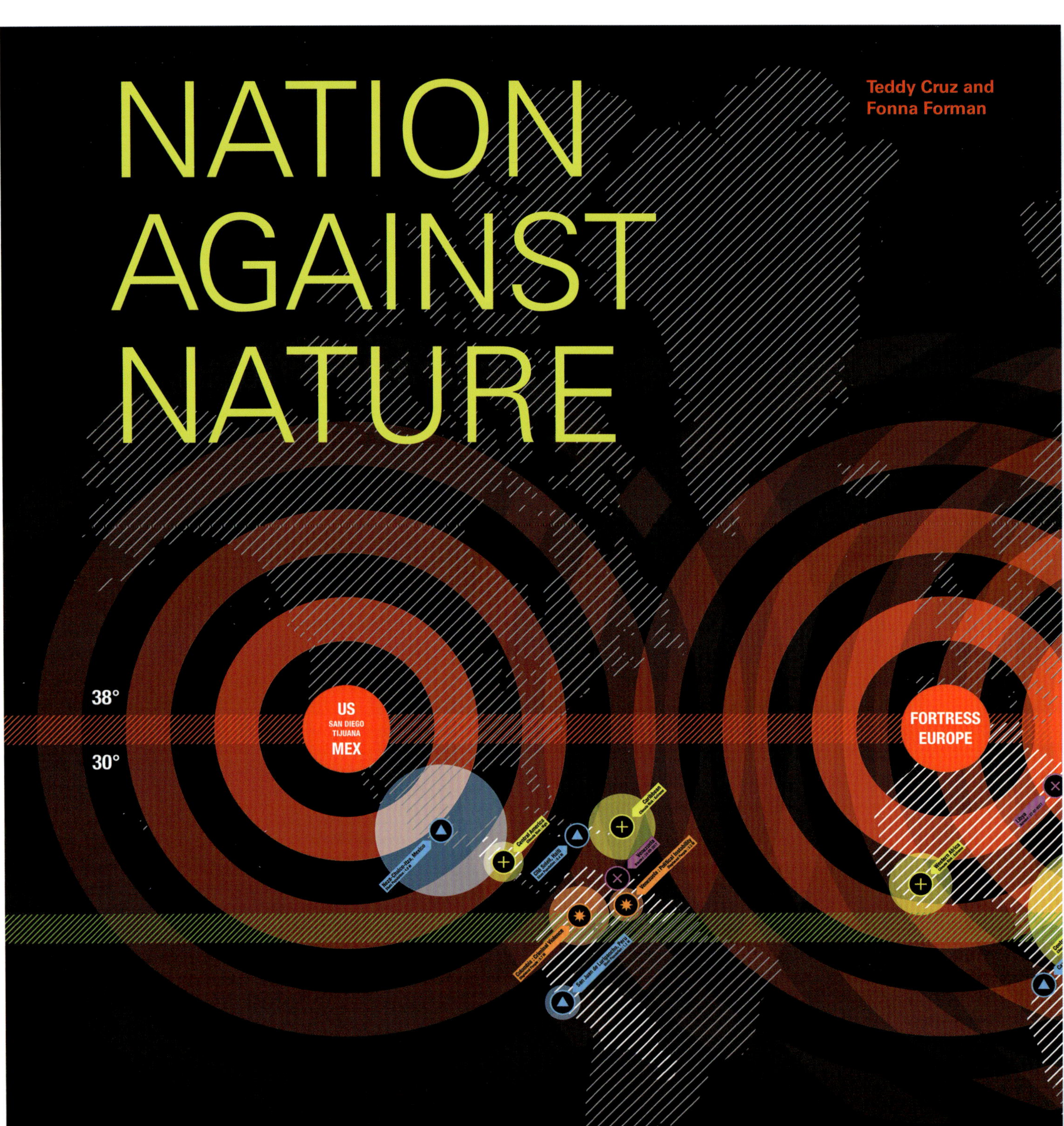

Teddy Cruz and **Fonna Forman** are professors at the University of California, San Diego. Here they discuss their practice investigations of the border conditions between Mexico and the US, proposing a Cross-Border Commons of political coalitions, ecological zones and new forms of citizenship.

FROM THE GLOBAL BORDER TO THE CROSS-BORDER COMMONS

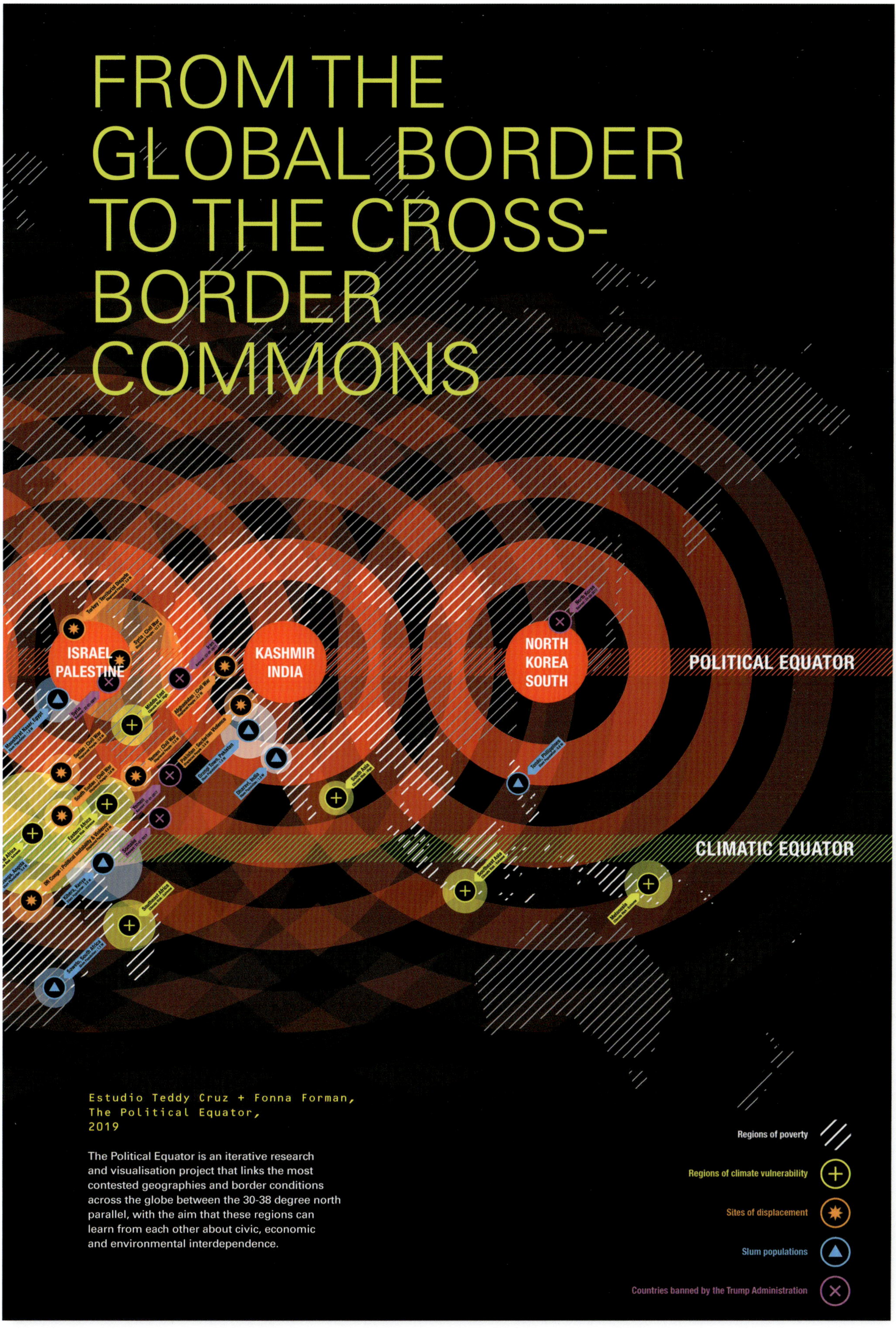

Estudio Teddy Cruz + Fonna Forman,
The Political Equator,
2019

The Political Equator is an iterative research and visualisation project that links the most contested geographies and border conditions across the globe between the 30-38 degree north parallel, with the aim that these regions can learn from each other about civic, economic and environmental interdependence.

Estudio Teddy Cruz + Fonna Forman is a research-based
political and architectural practice located at the US-Mexico
border, in the binational metropolis of San Diego/Tijuana.
In the current geopolitical climate of escalating tension and
militarisation, the practice has been calling for transgressive
experiments in 'unwalling' that enable people to see each
other anew and cultivate cross-border public commitments
towards a more inclusive, democratic and environmentally
connected binational region. Here, border zones become
laboratories for provoking a more speculative imaginary of
'cross-border citizenship'.

The Political Equator

Border regions across the world can learn from
each other about civic, economic and environmental
interdependence. The Political Equator (2012–19) is
an experimental visualisation project that traces an
imaginary line along the US-Mexico continental border
and extends it across a world atlas, forming a corridor of
global conflict between the 30th and 38th parallel north.
Along this imaginary border lie some of the world's most
contested thresholds, including the US-Mexico border
at San Diego/Tijuana, the main migration route from
Latin America into the US; the Strait of Gibraltar and the
Mediterranean, the main funnel of migration from North
Africa into 'Fortress Europe', the Israeli-Palestinian border
that divides the Middle East; India/Pakistan, a site of
intense and ongoing territorial conflict since the British
partition of India in 1947; and the border between North
and South Korea, emblematising Cold War tensions
forward to the present day.

When this global Political Equator is visualised alongside
the climatic equator, the convergence of geopolitical
borders, political marginalisation, climate disruption and
human displacement across the world becomes evident.

Of course, the real Political Equator is not a flat line.
While border walls are often conceived as physical
fortresses against the encroaching Global South, borders
are reproduced in peripheral neighbourhoods everywhere,
where public divestment, marginalisation, racism and
inequality divide communities and institutions. The
narratives of hatred and mistrust that circulate within and
outside of these geographies of conflict have been met with
irruptions of civic and political resistance, demanding new
and more inclusive imaginaries of coexistence.

The Nation against Nature

The continental border between the US and Mexico is
physicalised as a solid wall that interrupts the social,
economic and environmental ecologies of the region.
This jurisdictional line has incrementally hardened over
time. A chain-link fence in the 1970s, a steel wall constructed
with temporary landing mats discarded by the US military
after Operation Desert Storm in Iraq in the 1990s, and today
a see-through concrete pylon wall crowned by electrified
coils and panoptic night-vision cameras. The spectre of
Trump's 30-foot (9-metre) high continental wall reignites
worries about the delicate cross-border ecosystems it
violates, compromising the environmental health of human
communities on both sides.

Some of the eight borderwall prototypes built by the
Trump administration in 2017, viewed through a hole
in the existing wall.

Nation against Nature (2018) documents precise sites
along the continental border where the jurisdictional line
collides with natural systems. At these junctures, the
border wall disappears momentarily as the complexity of
topographic systems challenges the political artefact. A new
proposed border wall threatens to close these gaps, further
damaging transnational ecologies and harming both sides.
The work of Estudio Teddy Cruz + Fonna Forman provokes
a more ecological way of thinking about border spaces,
and a more inclusive idea of regional interdependence.

MEXUS: A Geography of Interdependence

MEXUS (2018) is a visualisation of the US-Mexico
continental border without the line, presented instead as a
transnational environmental zone comprising eight shared
watershed systems. By unwalling this thickened system
of interdependencies, MEXUS provokes a more inclusive
idea of citizenship based on coexistence, shared assets and
cooperative opportunities between divided communities.
The border is not a place where things simply 'end'.
MEXUS visualises what walls cannot contain: watersheds,
indigenous lands, ecological corridors and migratory
patterns. It becomes local and concrete at the Tijuana River
Watershed — at the precise juncture between the informal
settlement of Laureles Canyon and the Tijuana River Estuary.

While the US-Mexico border has been publicly maligned
as a site of violence and crime, division and fear, MEXUS
presents the national threshold as a zone of urban and
political creativity and experimentation, characterised by
bottom-up dynamics and invisible cross-border flows. It
rethinks the politics of identity and reimagines citizenship
beyond the jurisdictional limits of the nation towards a
more expansive idea of territory, grounded in shared
assets and opportunities.

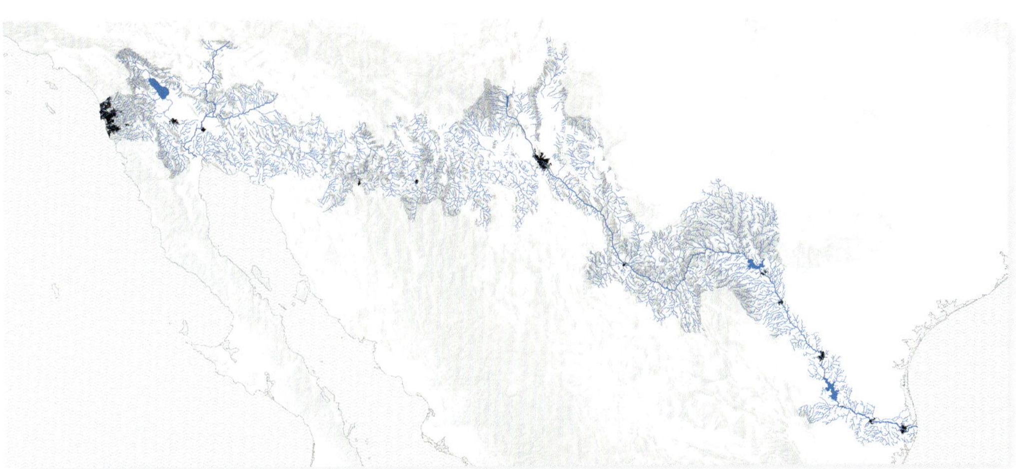

Regional Disruption: An Infrastructure of Insecurity

Zooming deeper into the westernmost watershed of MEXUS, one finds the Tijuana River bi-national watershed system, which is shared between the border cities of San Diego and Tijuana. Twenty-five per cent of this watershed is located in the US; 75 per cent is located in Mexico. The construction of new surveillance infrastructure by the US truncates the many canyons of this hydrological system. As they travel northbound, before discharging into the Pacific Ocean, these canyons must navigate the conflicts between a slum, the border wall, and a protected natural estuary.

This collision between natural and political forces is most profound at this juncture where Los Laureles Canyon in Tijuana, an important finger of the binational watershed, crosses the border line, and culminates in the Tijuana River Estuary in San Diego. This sensitive environmental zone flanking the border wall has been impacted in recent years by the activities of Homeland Security building a militarised 'third wall' and other infrastructures of surveillance and control. The informal Tijuana settlement of Los Laureles sits in one of these canyons, at a higher elevation than the Estuary. Construction of the new border wall has accelerated the northbound flow of waste from the slum into the estuary, syphoning tons of trash and sediment with each rainy season, and contaminating one of the most important environmental zones, the 'lungs', of the bio-region.

Although the wall is presented to the American public as an object of national security, it is actually a self-inflicted wound – the cause of great international environmental and economic insecurity in the years to come.

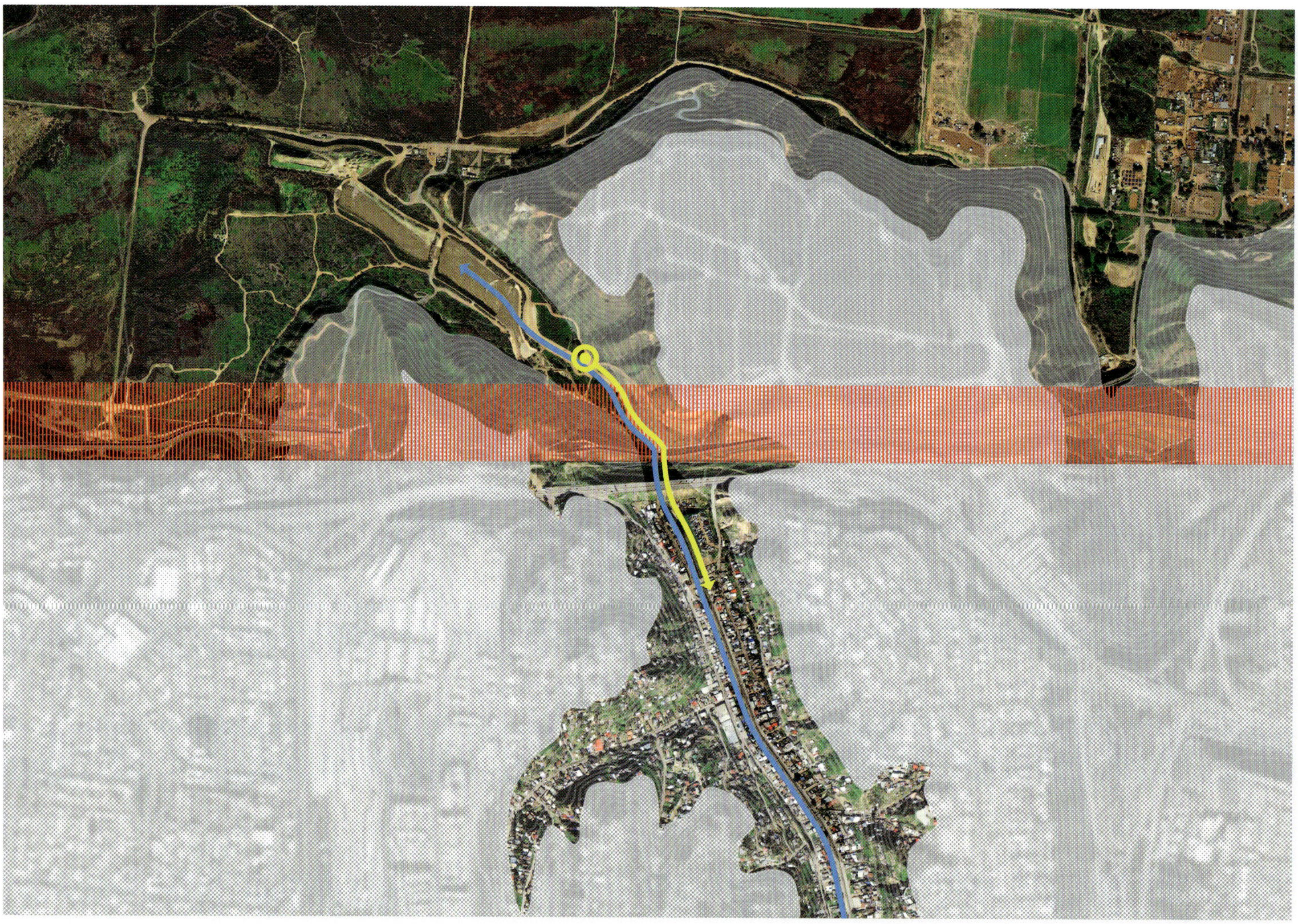

Estudio Teddy Cruz + Fonna Forman,
Border-Drain Crossing, Imperial Beach,
California and Tijuana, Mexico, 2011

The practice organised a nomadic public performance in 2011, in which 300 participants walked across the border southbound through a border drain in the wall, from a Federal US estuary into an informal Tijuana settlement, to raise public awareness of the environmental interdependencies between Tijuana and San Diego.

Estudio Teddy Cruz + Fonna Forman curated a cross-border public action through a sewage drain underneath a section of the border wall

Mexican border agents await to stamp the passports of participants on the Mexican side of the border-drain.

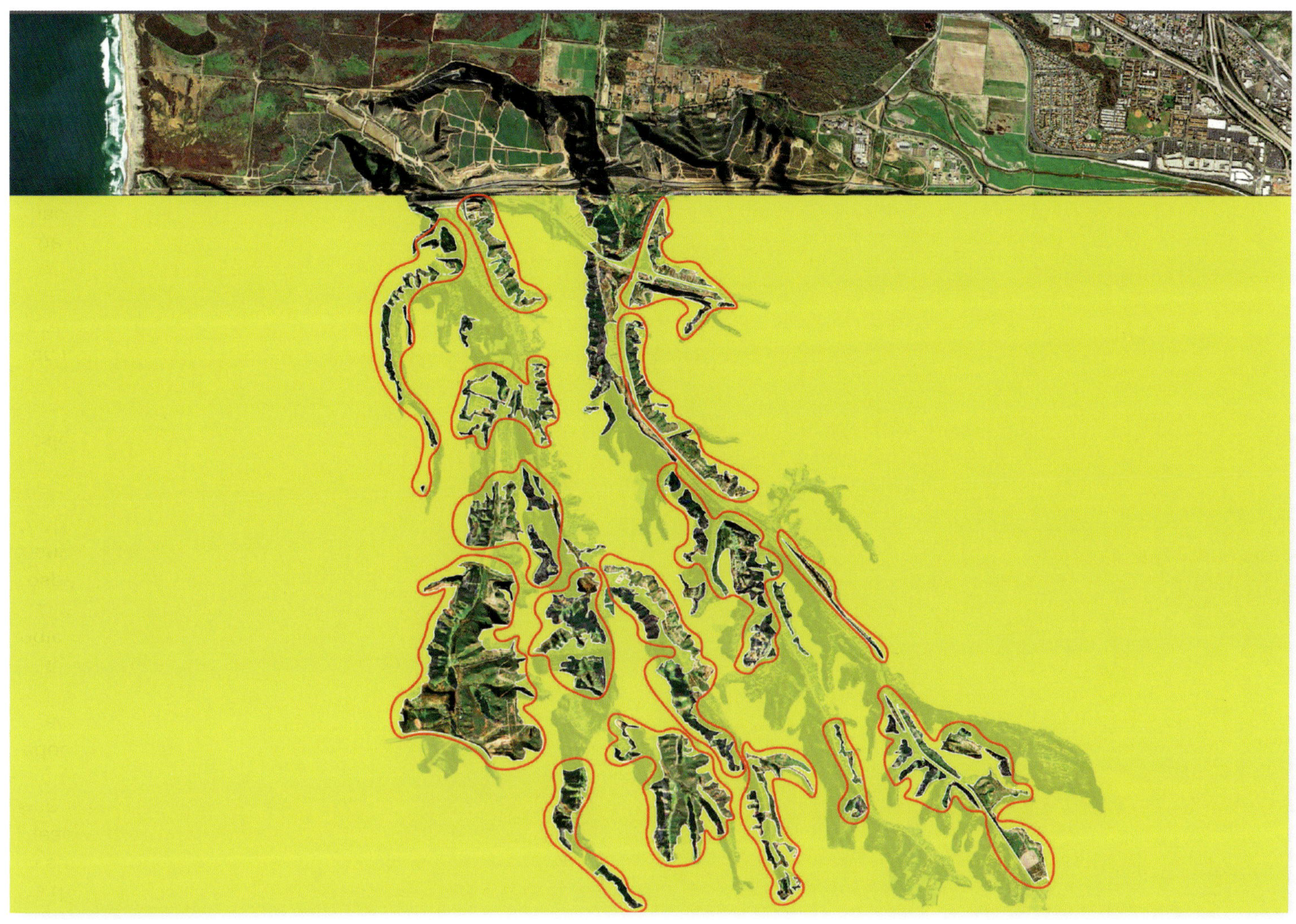

Estudio Teddy Cruz + Fonna Forman,
Cross-Border Commons,
2019–

A proposed binational land conservancy will link the informal settlement of Los Laureles Canyon in Tijuana with the Tijuana River Estuary in the US.

Border-Drain Crossing

In 2011, Estudio Teddy Cruz + Fonna Forman curated a cross-border public action through a sewage drain underneath a section of the border wall recently built by US Homeland Security, located at the precise point where the informal settlement in Mexico collides with the estuary on the US side. A permit was negotiated with Homeland Security to transform a drain under the wall into an official port of entry for 24 hours. They agreed, as long as Mexican immigration officials were waiting on the other side to stamp passports. As the 300 participants moved southbound under the wall against the natural northbound flow of slum wastewater headed towards the estuary, they reached Mexican immigration officers who had pitched a tent on the south side of the drain, inside Mexican territory. The strange juxtaposition of pollution seeping into the environmental zone, the stamping of passports inside this liminal space, and the passage from pristine estuary to slum under a militarised culvert, amplified the region's most profound contradictions and interdependencies.

Cross-Border Commons

Walls undermine vital regional ecosystems essential to the survival of the communities on both sides. Can a more just, cross-border public be mobilised to steward the shared environmental interests between two divided cities? Can border regions with shared environmental assets become laboratories to reimagine citizenship beyond the nation-state?

The practice is now leading an ambitious cross-border coalition of state and municipal governments, communities and universities to establish a Cross-Border Commons, a transgressive land conservancy that identifies slivers of land in the Mexican slum of Los Laureles, bundles them, and connects them with the Tijuana River Estuary on the US side. This 'green cross-border stitch' forms a continuous political, social and ecological zone that transgresses the international line. ⚮

Text © 2020 John Wiley & Sons Ltd. Images: pp 114-15, 117-19 © Estudio Teddy Cruz + Fonna Forman; p 116 © Estudio Teddy Cruz + Fonna Forman, photo Jona Maier.

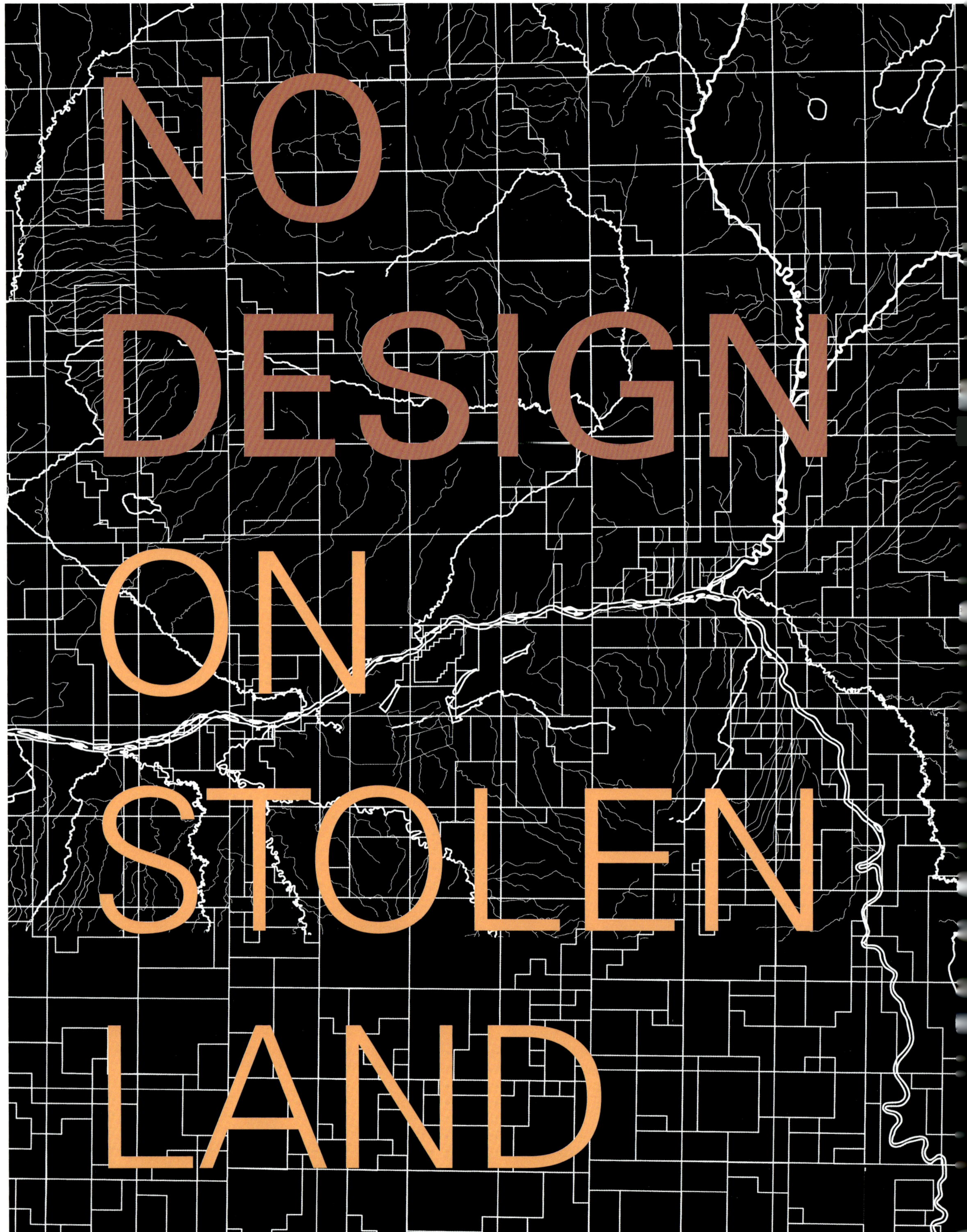

NO
DESIGN
ON
STOLEN
LAND

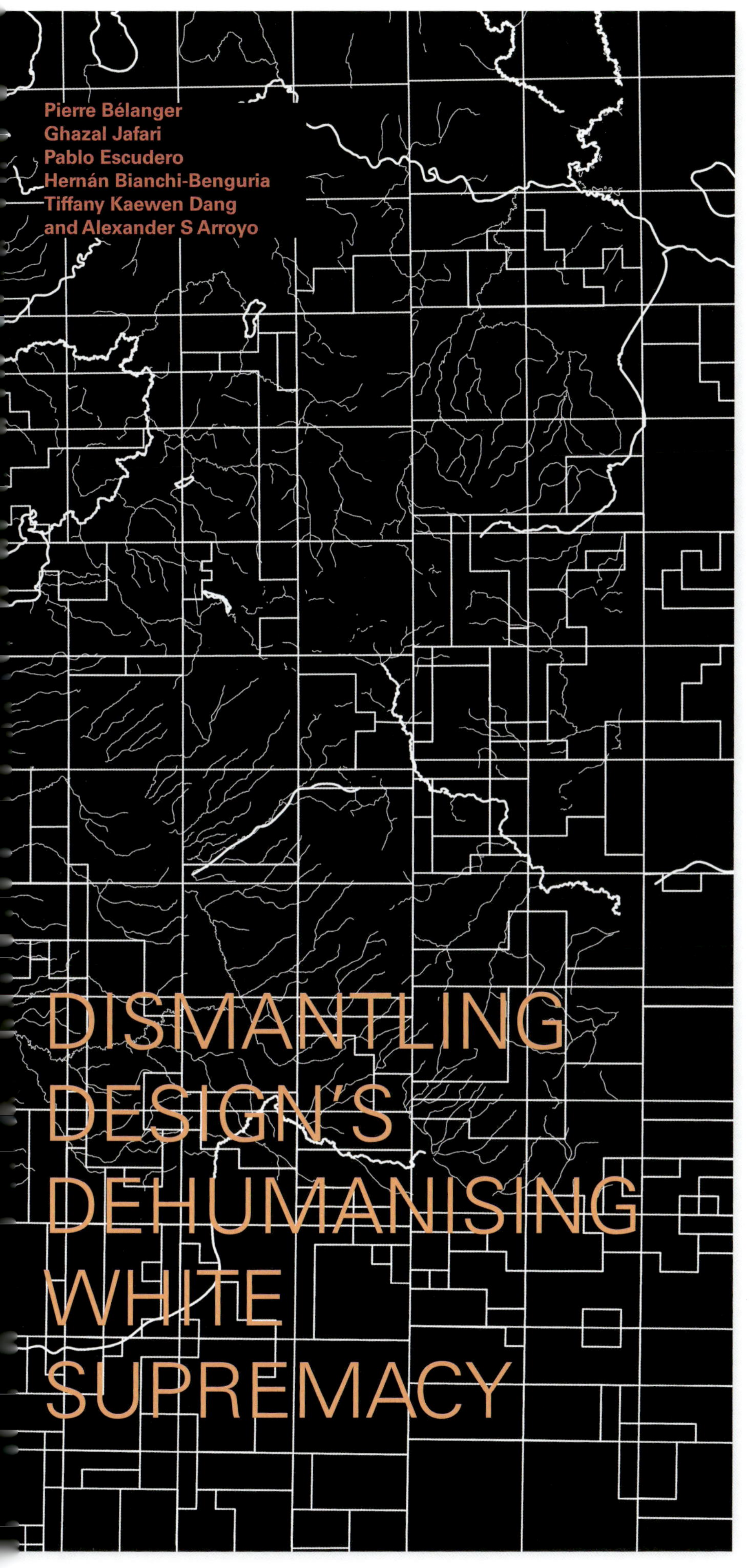

DISMANTLING DESIGN'S DEHUMANISING WHITE SUPREMACY

Capitalism is predicated on property ownership, and particularly land. The US and Canada are premier examples of the appropriation and dispossession of land occupied by Indigenous Peoples. **Pierre Bélanger** and co-founders at OPEN SYSTEMS (OPSYS) based in Boston, on traditional lands of the Massachusett Peoples, illustrate tactics of 'undesign' adopted in this context.

Oil Sand Lease Ownership,
Northern Alberta,
2016

Although seemingly invisible, the map of the Athabasca River and Treaty 8 Lands underlies the Quadrangles of Drilling Permits & Exploration Districts that are inscribed within the original 1870–1930 Dominion Land Survey System Grid. Data source: Terracon Géotechnique, 2018 Oil Sands Lease Map and Safety Schedule: www.terracon.ca/site/assets/files/3235/terracon-oil-sands-lease-map_2018-min.pdf.

Every single building site – from a house to a highway – benefits from the exploitation of a capitalist property regime built on the back of broken treaties. These sites are not only taken from stolen lands and unceded territories,[1] they are the spatial products of a violent structure and system of settler-colonialism that displaced and continue to dispossess Indigenous Peoples through 500 years of territorial injustices. *Mundus Novus. Terra Nullius. Doctrine of Discovery. Manifest Destiny.* Since 1492, this system of state policies of segregation, assimilation or extermination continues today, literally mining squandered lands and violating Indigenous sovereignties through forced removal, gendered violence, police brutality, cultural appropriation, underservicing and overincarceration.[2] As genocide,[3] the oppressive system of settler colonialism is now normalised through contemporary urbanism, precisely because of the 'denigration of Indigenous culture. Basically, it's racism … systemic, institutional, individual, interpersonal racism.'[4] As Deborah A Miranda writes in 'Teaching on Stolen Ground' (2007), 'genocide depends upon the appropriation of the identity of the colonized by the colonizer. Misinterpretations and misrepresentations of Native culture, religion, character, and worldview for consumption by the nonindigenous are the crucial elements in such a genocidal agenda.'[5] Inalienable Indigenous rights denied by privileged descendants of Christian Europeans – either by ignorance, neglect or design.

Design Dehumanises

Infrastructural systems are not only planned, engineered and built on stolen lands, they are codified as state systems of erasure that lend the appearance of permanence. They dishonour original treaties by denying the basic principles of consent and community consultation. Masterplanned

Depiction of the cultural genocide and violent extermination of Indigenous Peoples (including women, children and Two-Spirit) as well as physical abuse of African slaves by Spanish colonising forces; early preconditions and precursors to settlement and urban civilisation of Central and Southern America in the mid-16th century.

forgetting and bureaucratic stonewalling, set in and between the lines of colonial law and settler rule. 'Whether realized or implied, physical violence is important to the legitimation, foundation, and operation of a Western property regime. Certain spatializations, notably those of the frontier, the survey, and the grid, play a practical and ideological role at all these moments.'[6] Suspiciously 'under-researched',[7] in the paper world of design – maps, plans, codes, graphics – urbanism *is* colonialism. Upheld by colonial-era constitutional rights, the plunder from looted landscapes has left behind legacies of broken bodies and fractured families across toxic terrains of endless extractivism, a 'non-reciprocal, dominance-based relationship with the earth' that Naomi Klein identifies as 'one of purely taking'.[8]

Design Dispossesses

They call it 'design' for a reason. Destruction of signs, signals, symbols, signifiers – oral, traditional, cultural, political, territorial. As a substitute for signs, place names subvert spatial identities. From Alaska to Oklahoma, racist toponomies are the white man's corruption of original places and peoples, perpetuated by what Black Canadian Studies scholar Charmaine Nelson identifies as the 'colonial, cartographic imagination'.[9] Design not only destroys and distorts memory, it ruptures tradition by imposing techno-scientific terms and transplanting state names that edify white supremacy through Cartesian ideology. Mapmakers as liars. Land use is rationalised terrain, by racialisation. From waste colonialism in New Mexico on Navajo Lands to industrial capitalism in Amazonia throughout the territories of Quichua Peoples, to resource extractivism in Alberta and Saskatchewan on lands and waters of Dene, Cree and Métis Peoples. Design aids and abets by obeying jurisdictional powers that rule over everything from deserts, forests and rivers to streets, parks and sewers. Territorial planning, structural engineering and building specifications are its arms. The genius loci of dumps, dams, pits, pipes and mines are its monuments. Radium, mercury, arsenic, lead, phosphorus, ammonia its poisons.

Design Whitewashes

Laundering land has a long history. Planners, engineers, architects, illustrators are its bleaching agents. The deep geological repository for the 10,000-year storage of transuranic radioactive waste on ancestral lands of the Navajo Peoples, operated today by the multinational engineering conglomerate AECOM. Natural history's 10,000-year-old Willamette Meteorite stolen from the Clackamas tribe of Oregon, moved to New York City's American Museum of Natural History in 1906, then relocated to its Rose Center for Earth & Space, designed by architects James Polshek and Todd Schliemann in 2000. The horrific story of Matoaka violently taken away from the Powhatan people in the 17th century by English colonisers then romanticised in Disney's 1995 *Pocahontas*, directed by Mike Gabriel and Eric Goldberg. Jeff Bezos's trademarking of the world's longest river in 1994 into the world's largest transnational corporation, branded with a smug smile by Turner Duckworth designer Anthony Biles. Senator Elizabeth Warren's appropriation and claims to Cherokee identity in 1986, now running for the 2020 presidential election. Lies on the land.

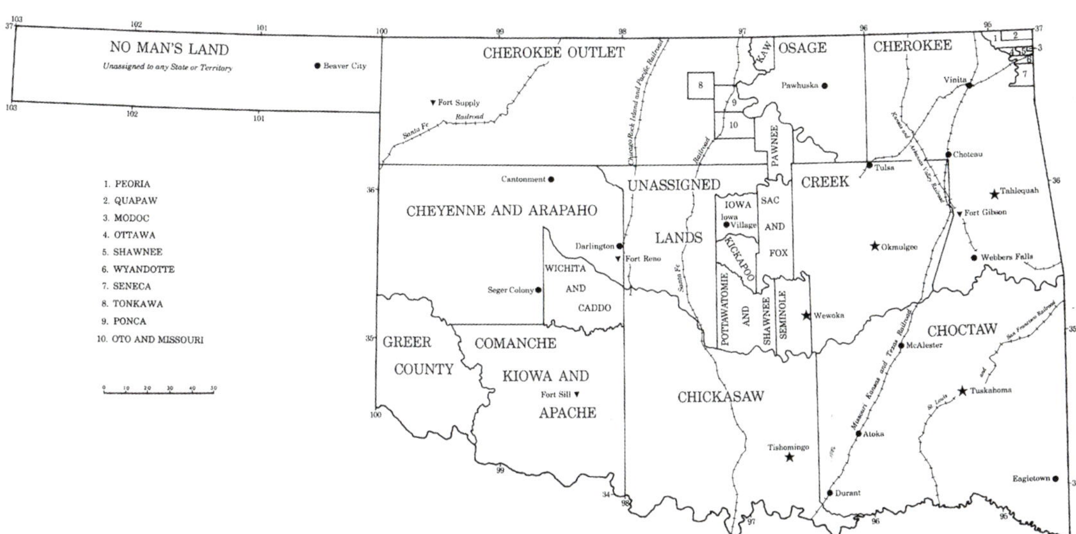

Map of Indian territory
and No-Man's Land,
Atlas of Oklahoma,
1866–89

Delineating displacement by
design, the outlines of the State of
Oklahoma following the 1830 Indian
Removal Act and the 1887 Dawes
Act to Indian Territory, where nearly
40 different Native American tribes
were relocated from across Turtle
Island (North America).

James Polshek and Todd Schliemann,
Rose Center for Earth & Space,
American Museum of Natural History,
New York,
2000

Located on the lower level of the Dorothy and Lewis
B Cullman Hall of the Universe, the 10,000-year-old
Willamette Meteorite was stolen from the Clackamas
tribe of Oregon and moved to New York City's
American Museum of Natural History in 1906.

Bunky Echo-Hawk,
Not Your Mascot,
2016

Contesting baseball's racist Chief Wahoo
mascot unapologetically worn by the
Cleveland Indians since 1946, designed by then
17-year-old draughtsman Walter Goldbach, and
the Washington Redskins logo dating back to
the 1932 Boston Braves NFL football team.

Design Alienates and Excludes

Representing a growing 5 per cent of the world's population according to the United Nations, Indigenous Peoples live on, care for and nurture a majority of the lands, skies and waters of the planet.[10] Yet, city and territory are strategically segregated by the ever-increasing divide between spaces of consumption and production; lands fragmented by logistical landscapes. Cut up by tight metropolitan grids, littered with architectural effigies, spreading out into settler suburbs, lands are crisscrossed and waters butchered by hard technological infrastructures and 'fine-grained spatial technologies of power' policed by soft regulatory systems of zoning bylaws, procurement regulations and building codes.[11] This dominant matrix not only leaves little room for the body politic of the under-represented communities of Black, Indigenous and People of Colour who live and work in cities today; the concrete realities of the metropolis serve as a brutal reminder of colonisation's accumulating monumentality. Designers as contractors. The General Lee Monument in New Orleans: confederate-era edifice designed as settler-colonial victory trophy by sculptor Alexander Doyle and architect John Roy in 1884. The St Louis Arch on the Missouri River: frontier gateway to the American West designed by Finnish-American architect Eero Saarinen in 1935. The Mount Rushmore, in Keystone, South Dakota, on the sacred Black Hills lands of Lakota Sioux Peoples, conceived by Doane Robinson and carved by Ku Klux Klan sympathiser Gutzon Borglum between 1927 and 1941. Central Park in Manhattan, New York: African-American farms on former Lenape territory cleared for bourgeois open space by landscape architect Frederick Law Olmsted and architect Calvert Vaux in 1857. Yellowstone in northwest Wyoming, on lands of Indigenous Peoples of the Great Plains, a national public park linked to a system championed by conservationist John Muir from 1872 onwards.

Settler urbanism imposes a spatial code on the oppressed, reduced to users and consumers

Doane Robinson and Gutzon Borglum,
Mount Rushmore National Memorial,
Keystone,
South Dakota,
1927–41

The monumentality of white supremacy in US National Monuments was controversially carved into the sacred Black Hills, depicting four white, male American Presidents. It has been the subject of countless protests from the Great Sioux Nation since its creation, in the 1970s, and more recently in 2018.

Design Subjugates

From towering complexes in the colonial metropolis, the darkness of settler urbanism casts long shadows on surrounding resource hinterlands. Its high priests, a lineage of privileged historians, call it 'the city' where more than half of the world's population now apparently lives. Settler urbanism imposes a spatial code on the oppressed, reduced to users and consumers, in broad daylight using a geographic information system of distanced surveillance and remote control across an array of different elevations, altitudes, atmospheres. Controlled airspace. Plantation logic.[12] From Sigfried Giedion to Alan Gowans, Lewis Mumford to Henri Lefebvre, white men have preached the gospel of settlements ad nauseam while chronicling its many frontiers from academic outposts, rarely setting foot on the ground. Best qualified as 'urbs nullius', design of white space as sanctioned gentrification.[13]

Design Masks

For every city, there is a treaty or a title (broken or not), whether the land (unceded or not) was taken by war, invasion or surrender. So, there is nothing novel or revolutionary about urbanism. Only tricks, traitors, terrorists.[14] Broken treaties make cities a war zone of occupied territories. Urban dwellers, space invaders – the hostiles. Washington DC is on ancestral lands of Anacostank, Piscataway and Pamunkey Peoples. Ottawa, lands of the Omàmiwininìwag Peoples. New York City, lands of the Lenape Peoples of the Delaware nation. Montreal, lands of the Kanien'kehá:ka Nation. Los Angeles, lands of the Gabrielino, Tongva and Tataviam nations. Santiago, lands of the Mapuche Peoples. Rio de Janeiro, lands of the Tupi and Guaraní Peoples. If 'we are all Treaty Peoples,'[15] then 'the evidence of our betrayal lies all around us … we [settlers] have not lived up to our end of the bargain,' because 'it takes two to make a treaty'.[16]

Design Launders

The art world is no exception to design's delusions. Settler artwork (especially European) precisely depends on physical erasure of Indigenous bodies and political suppression of sovereignties to fabricate remoteness and emptiness – modern settler space. Enter the anti-establishment of land artists from the 1970s, founders of the earthworks movement in the so-called middle of nowhere. Michael Heizer's *Double Negative* (1969) carved out of ancestral lands of the Southern Paiute Peoples. Robert Smithson's *Spiral Jetty* (1970) and Nancy Holt's *Sun Tunnels* (1973) built on ancestral lands of Ute, Diné (Navajo), Paiute, Goshute and Shoshone Peoples. Walter de Maria's *Lightning Field* (1977) on traditional territories of Apache and Navajo Nations. Illegitimate earthworks rife with art-historical traditions that exploit the displaced and dispossessed since the 18th century. Federally disposed lands by the US Department of Interior, spread out across a Cold War laboratory landscape of military bases, testing sites and dumping grounds in the headwaters of Native American reservations. Twentieth-century con artists, complicit in the duplicitous contamination of Indigenous bodies. No justice, nowhere.

Official Seal of the City of New York, 2019

Since its original creation by the Dutch in 1626, the official mark of New York City has been revised, reconceived and redrawn into more than a dozen versions by both British colonial agents and American settlers. However, the entrenchment of the oppressive symbolism of settler urbanism and semiotics of stolen land is persistent.

Anonymous,
Occupation at Alcatraz Island,
San Francisco Bay,
California, 1969

As a declaration of sovereignty over lands stolen by the federal government, the 78 Indigenous members of the group calling themselves Indians of All Tribes took part in a 19-month standoff at the site of the former federal prison site that later became part of the National Park System as the Golden Gate National Recreation Area in 1972.

Design Destroys

In this settler-colonial plot, conservation is the con job that normalises extinction and naturalises domination. For the naturalist junta – the generation of conservationists from Theodore Roosevelt to Henry David Thoreau – preservation of nature was the preservation of white supremacy. Naturalism of nationalism. Conservationism as heroism. The picturesque, an imperial gaze. Obsessed with eurocentric maps of New World–Old Word divisions, missionary men like George Perkins Marsh have manhandled nature, as in his 1864 biblical *Man & Nature*. Heteropatriachal man, conveniently universalised to consume the 'New World' by disarming Indigenous systems of governance and kinships to rearm a 'wilderness' for 'the civilised' and 'the superior', free of 'savages' or 'Blacks' recounted in racist, frontier-fantasy novels. Ernest T Seton's 1912 *The Book of Woodcraft & Indian Lore*, fake lore. For naturalists, conservationists, foresters, ecologists and even presidents, wilderness weaponised as 'the raw material out of which man hammered the artifact called civilization'.[17] Predators.

Settler-Colonial Statecraft

The Christian wrecking ball that cleared forests not only made way for an imagined wilderness, but like Seton's 1903 fiction *Two Little Savages* was repopulated by young boys playing cowboys and pretend Indians. Colonial tropes re-enacted, primitive living invoked, extinction narratives replayed. Riding the coat-tail of George Catlin's ethnographies of colonial encounters and paintings of American exploits, the British imperialist Robert Baden-Powell epitomised colonial conservationist ideologies after escapades in Rhodesia and South Africa; scouting for boys since 1907 across the Commonwealth from Australia to Canada. His mission led to the foundation of the Boy Scouts of America, a worldwide brotherhood of scouts and volunteers now 30-million-strong. Pledging allegiance to 'God, Country, and Self', its divine ethos is disguised as patriotic oath: the revival of colonial-era heteropatriarchy fuelling a white male saviour complex formulated precisely when the British Empire was falling. Scoutcraft as masquerade for statecraft.

So, what if the design world responded to what Leah-Simone Bowen and Falen Johnson identify in *The Secret Life of Canada* as the 'conditioning of active unawareness'[18] by honouring the historic treaties and agreements drafted over centuries that bind us all? Could that ensure a ground to live on and a future to fight for in the next five centuries?

Surrender

This counter-design claim confronts the inseparability of climate change embedded in racialised urban spaces and capitalist structures of settler-colonialism that are upheld by a dwindling white majority. The claim is conceived for settler designers and scholars in reaction to the near-total apathy, ignorance, erasure and marginalisation of the identities, genders, histories, territories and rights of Black, Indigenous and People of Colour that form a complex, conflicted and entangled web of urban and territorial ecologies upon which the professional disciplines of design intervene. Counter-representation entails concessions and crossovers, a powerful kind of disciplinary transformation that Rod Barnett aptly qualifies in 'Designing Indian Country' (2016) as 'intersectional discourses of race and ethnicity, sexuality, and materialism that have challenged self-conceptions [of scientific or professional practices] and opened up more radical modes of practice'.[19]

Rereading

To do this, the system and structures of settler urbanism, including the technocratic standards of design disciplines and spatial orders of projects, need to fundamentally change. This requires a rereading of treaties and understanding of their inherent anti-colonial principles to honour their responsibilities and embody relations because bodies cannot be separated from their territories. This *is* the project of un-design: a spatial language and way of working to overwrite the present by retroactively underscoring the past. There can be no design without the process of decolonisation.

Rob Wilson,
Honor the Treaties – spray-painted on a segment of the Line 3 pipeline in Superior, WI,
2017

Passing through lands of the 1855 Treaty of Washington, the Line 3 pipeline project illegally crosses near or through several Indian reservations including White Earth, Red Lake and Leech Lake, home to extensive regions of wild rice lakes in the headwaters of the Mississippi River. Line 3 is a 1,600-kilometre (1000-mile) crude-oil pipeline operated by Canadian resource distribution corporation Enbridge Inc, connecting the Tar Sands in Alberta to oil refineries in Wisconsin.

Transgression

To undress this carceral landscape requires the unmapping of settler urbanism. It means destroying the dispossessive categories that sanction exclusion, exploitation, extraction and erasure. Dismantling the structures that obviate the legal landscape of treaties and that are constructed to sever relations between lands, waters, beings, cycles and communities. Unplanning oppressive policies. Unnaming colonial place names. Debasing base maps. Debunking benchmarks. Redrawing legends. Retroceding lands. This démontage bends rules and rewrites settler-colonial code. If it does not break the law, it is not new.

Unbuilding

The geopolitical system of settler colonialism that continues to erase Indigenous voices and marginalise Indigenous land rights must be confronted. As Clayton Thomas-Müller called for in 2016, 'change the system, not the climate'.[20] Essential here is the unconditional adoption and enactment of 'Free, Prior, Informed Consent' from Indigenous nations based on the 2007 United Nations Declaration on the Rights of Indigenous Peoples as a protocol of humane practice before any plans for transformation take place on the ground.

Whose lands are you on? Which territorial treaties are they part of? Who are you accountable to? Whose stories and histories are privileged? Who are your collaborators? Are waters, rivers, estuaries, streams, seedlings, beavers, and other beings part of that change? These concluding questions may seem extraordinarily banal, but yet are essential in unbuilding the structures of settler urbanism and weakening the systems of whiteness that have destroyed so much. To ground territories with their treaties, this means renewing relations, building alliances and embodying anticolonial measures. Until design becomes a ghost of its dehumanising self, the project of dismantling its true oppressive nature will never end. ⌀

Notes

1. 'Stolen land and stolen labor are the essential requirements of capitalism.' See Owen Toews, *Stolen City: Racial Capitalism and the Making of Winnipeg*, ARP Books (Winnipeg), 2018, p 18.
2. Patrick Wolfe, 'Settler Colonialism and the Elimination of the Native', *Journal of Genocide Research*, 8 (4), 2006, pp 387–409.
3. Justice Murray Sinclair, 'Statement from the Chair', *Truth & Reconciliation Commission of Canada I*, 2015, p 4.
4. Raven Sinclair in conversation with Connie Walker, 'Missing and Murdered: Finding Cleo', CBC News Podcast, Season 2, Episode 8, 20 March 2018.
5. Deborah A Miranda, 'Teaching on Stolen Ground', in Jennifer Sinor and Rona Kaufman (eds), *Placing the Academy: Essays on Landscape, Work, and Identity*, Utah State University Press (Logan, UT), 2007, p 181.
6. Nicholas Blomley, 'Law, Property, and the Geography of Violence: The Frontier, the Survey, and the Grid', *Annals of the American Association of Geographers*, 93 (1), March 2003, p 121.
7. Anthony D King, *Urbanism, Colonialism, and the World-Economy*, Routledge (London), 1990, p 2.
8. Naomi Klein, *This Changes Everything*, Simon & Schuster (New York), 2014, p 169.
9. Charmaine Nelson, 'Interrogating the Colonial Cartographic Imagination', *American Art*, 31 (2), Summer 2017, pp 51–3.
10. Victoria Tauli-Corpuz, 'Attacks Against and Criminalization of Indigenous Peoples Defending Their Rights', Report to Human Rights Council, 39/17, United Nations, 2018: http://unsr.vtaulicorpuz.org/site/index.php/en/documents/annual-reports/251-report-hrc2018.
11. Jane M Jacobs, *Edge of Empire: Postcolonialism and the City*, Routledge (London), 1996, p 21.
12. Katherine McKittrick, 'On Plantations, Prisons, and a Black Sense of Place', *Social & Cultural Geography*, 12 (8), 2011, p 951.
13. Glen S Coulthard, *Red Skin White Masks: Rejecting the Colonial Politics of Recognition*, University of Minnesota Press (Minneapolis, MN), 2014, p 264.
14. Alanis Obomsawin, *Trick or Treaty*, National Film Board of Canada, 2014: 84 mins.
15. Tara Williamson, 'We Are All Treaty Peoples', *Decolonization: Indigeneity, Education & Society*, 24 December 2012: https://decolonization.wordpress.com/2012/12/24/we-are-all-treaty-people/.
16. Adrienne Clarkson, 'The Society of Difference'/La Société de la Différence' *8th Annual LaFontaine-Baldwin Lecture*, 2007, p 28: www.icc-icc.ca/site/site/uploads/2016/11/LaFontaineBaldwinLecture2007_AdrienneClarkson.pdf.
17. Aldo S Leopold, *A Sand County Almanac: With Essays on Conservation from the Round River*, Oxford University Press (New York), 1949, p 264.
18. Leah-Simone Bowen and Falen Johnson, 'The Secret Life of Banff', The Secret Life of Canada Podcast, Episode 1, 31 August 2017: https://passport2017.ca/articles/episode-1-secret-life-banff
19. Rod Barnett, 'Designing Indian Country', *PLACES*, October 2016: https://placesjournal.org/article/designing-indian-country/.
20. Clayton Thomas-Müller, 'Change the System, Not the Climate', World Social Forum, Montreal, 9–16 August 2016: https://m.youtube.com/watch?v=rlHzhlZoyk8.

To ground territories with their treaties, this means renewing relations, building alliances and embodying anti-colonial measures. Until design becomes a ghost of its dehumanising self, the project of dismantling its true oppressive nature will never end

St Alfege:
Hawksmoor Speaking Across Time

A Word from
Δ Editor Neil Spiller

**Simon Withers,
3D scans of St Alfege Church,
Greenwich, London,
2019**

opposite: South elevation. Scanning has revealed a whole host of
Hawksmoorian spatial relationships not seen or imagined before,
particularly between the crypt and the main body of the church.

... the art of Shaddowes you must know well Walter ... It is only the Darkness that can give trew Forme to our Work and trew Perspective to our Fabrick, for there is no Light without Darkness and no Substance without Shaddowe.
— Peter Ackroyd, *Hawksmoor*, 1985[1]

This issue of ⏀ is about landscape and widening its definition to include all manner of demographic, political, ethical and technological issues that constitute the contemporary world. There are other landscapes too. Buildings are a subset of landscape, and the work across a good architect's career is an evolving landscape of ideas, forms and beliefs that speak over time. As time goes on, this lexicon grows into a filigree of influences on others, some using the work as precedents for more modern buildings, others as a source for academic research, and others still as inspiration for fictional story-telling. Nicholas Hawksmoor (1661–1736) has provoked more than his fair share of this lacework of latter-day associations.

Hawksmoor is a prime example of social mobility. Born on to a yeoman farm in Nottinghamshire, he became a young clerk to a judge and later to Sir Christopher Wren, working his way up to a number of prestigious and influential positions throughout an eventful career. He is now regarded as one of, if not the greatest master of English Baroque architecture. Unlike many of his contemporaries, he did not enjoy the luxuries of the Grand Tour of Europe's great architecture; his knowledge of architecture was thus gleaned from engravings of Roman, early Christian and Ancient buildings and a voracious appetite for books.

This may account, in part, for his architectural originality. One can almost refer to him as the first Punk architect; he rejected the classical rules and in so doing developed a personal language of architecture still rich with quotations, but with a whole new syntax – a poet in stone, rich in light and shadow played out across highly modulated forms and profiles. His architecture speaks to us across time.

In 1711, an Act of Parliament was passed for the building of Fifty New Churches in the centre of London and its outskirts. A Church Commission was formed, and appointed Hawksmoor as one of its two surveyors. Hawksmoor designed six churches outright of the eventually built 12. St Alfege was the first, built between 1712 and 1714. The other five were Christ Church Spitalfields (1714–29), St George-in-the-East (1714–29), St Anne Limehouse (1714–30), St Mary Woolnoth (1716–24) and St George Bloomsbury (1716–31).

Nicholas Hawksmoor,
St Alfege Church,
Greenwich, London,
1712–14

right: South elevation: east side. Hawksmoor's architectural lexicon tore up the established rules of classicism, developing a new classical language of bold lines and deep shadow.

The Beast Beneath

During the 1970s, these six churches attracted the attention and fermented the imagination of writer Iain Sinclair. At this time many of them were presiding over derelict and bomb-damaged districts and were poorly maintained themselves. Sinclair was working as a council gardener to support his as yet not successful writing career. One of the landscapes he got to garden was St Anne's Limehouse, and this is how he got to know, close up, the work of Hawksmoor. St Anne's, he felt, was a brooding place, of shadowy, magnetic, dark ways.

As architectural writer, historian and curator Owen Hopkins writes: 'Sinclair's impressions of Hawksmoor's churches were shaped by literally digging down into the very matter – the mud, filth and detritus – of East London. Beneath the jagged surface of London's past, Sinclair could not help but find layer upon layer of myth, intrigue and forgotten history.'[2] St Anne's, Sinclair surmised, in his fevered, psycho-geographic imagination, was the epicentre of a network of Hawksmoor's churches that formed a series of malevolent, psychic connections and were attractors for heinous events over time. In 1975, Sinclair's *Lud Heat*[3] was published, a well-crafted work of fiction that bent historical facts to 'prove' his intuitions and the existence of a dark side to Hawksmoor.

In the 1980s, Peter Ackroyd channelled *Lud Heat* to inspire his novel *Hawksmoor*,[4] which features a modern-day detective of the same name who is investigating a series of murders in the vicinity of the churches, constructed by a character called Dyer – an architect of the real Hawksmoor's time. The novel portrays the churches as beacons of black magic, sacrifice and murder. As it progresses, it condenses these parallel times into one time, actions and reactions leak into the other time and vice versa, and finally, the evil Dyer gains control over the detective. Though purely fictitious, again the theme of Hawksmoor speaking to us across the centuries endures.

Hawksmoor's first great church, which is the genesis for all the subsequent ones, St Alfege sits like a ship in the 'bay' of maritime Greenwich in Southeast London. It gives the impression of a great stone monolith that has just risen out of the ancient earth. The sun's movement creates changing vistas of deep shadows, bringing into relief its architectural lineaments. The stone itself has a whiteness that seems to emanate sunlight caught within it over centuries. The overall effect of these phenomena is reminiscent of the opening quotation to this article – a fictional remark uttered by Dyer in Ackroyd's *Hawksmoor*, but architecturally true nonetheless.

Delving into the Heart of Hawksmoor

Thanks to the National Lottery Heritage Fund, Greenwich has recently been delving into its own history, and Hawksmoor and others connected with the church have again communicated with us across time. The area has benefited from three awards: for the cleaning and restoration of the Painted Hall ceiling at the Old Royal Naval College and works to the hall's undercroft; for the cleaning, remedial and improved accessibility works to St Alfege as part of its 'Heart of Greenwich – Place and People' project; and finally for improvements in Greenwich Park – all of which are enabling new discoveries to be made.

The Painted Hall is often described as England's Sistine Chapel, and is one of the most spectacular and important baroque interiors in Europe. Its ceiling and wall decorations are by the British artist Sir James Thornhill, and were completed in 1707 and 1726. The hall is simultaneously allegorical, mythic and contemporary for its time. Hawksmoor designed the building in which it is housed. Hawksmoor's and Thornhill's paths also intersect in a more modest way at St Alfege, as both worked on the church as well.

James Thornhill,
Sketch for chancel painting,
St Alfege Church,
Greenwich, London,
date unknown

The discovery of this sketch further consolidates the relationship between Thornhill and Hawksmoor at Greenwich.

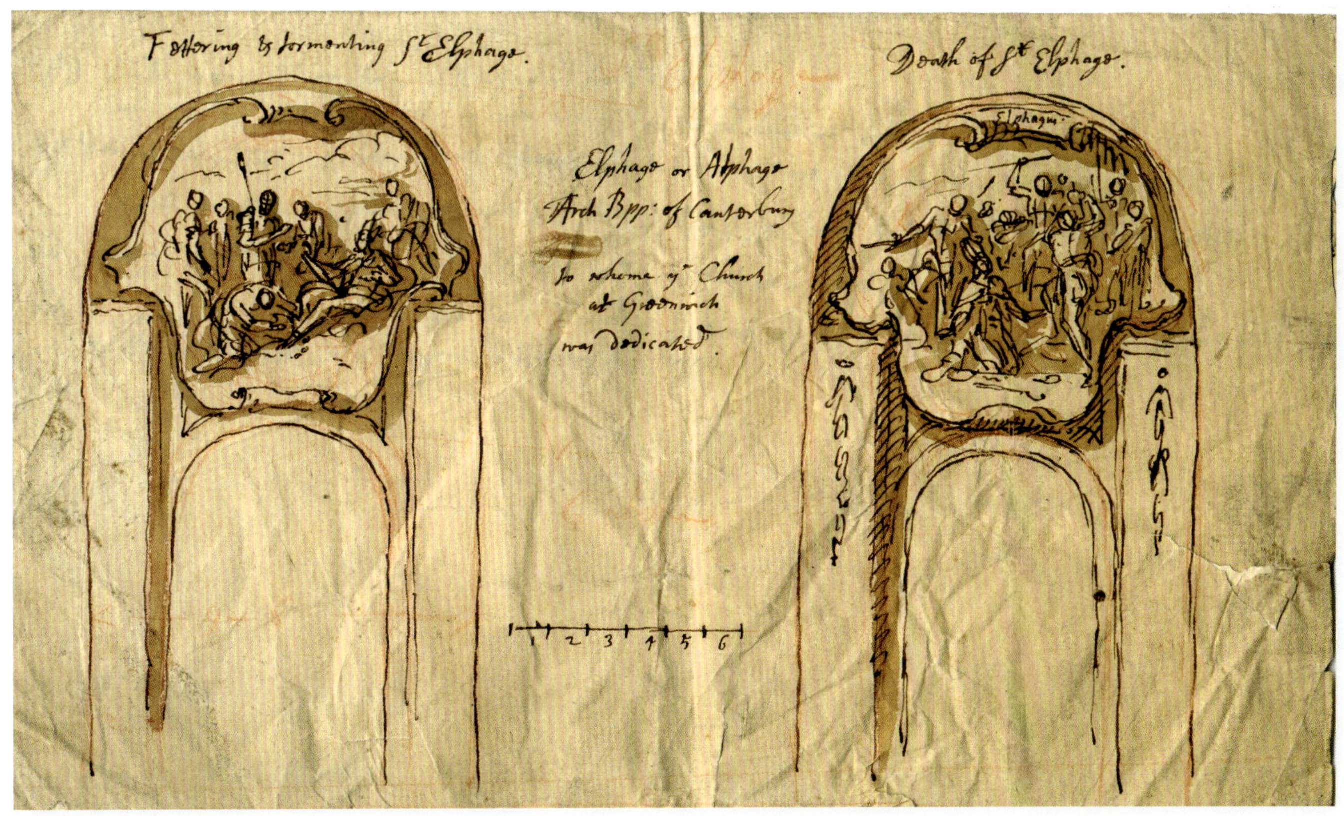

In 2018, Rebecca Parrant, the church's Heritage Engagement and Interpretation Manager, came across a sketch by Thornhill inside an envelope of unrelated, historical material. Parrant was accompanied by researcher Alison Fisher who instantly recognised the work. Thornhill's drawing appears to depict the story of St Alphege, Archbishop of Canterbury, who was murdered by the Danes in 1012 AD on the site of the church. The left part of the drawing is titled 'Fettering & tormenting St Elphage', and the right 'Death of St Elphage'. Fisher describes the content of the sketch thus: 'Within the drawing two sketches appear side by side and both show a composition that comprises an arch supported by two flat pillars. Initial research suggests that these might have been early concept proposals for the chancel painting at St Alfege Church. The existing painting in the chancel has been attributed to Thornhill's workshop.'[5]

Another significant discovery, in 2017, is a previously unknown Hawksmoor drawing of the north elevation of St Alfege, found in a box of old photos and news clippings at the Greenwich Heritage Centre by Richard Hill, of Richard Griffiths Architects, conservation architect for the 'Heart of Greenwich' project.

Across Time and Space
Also, as part of the project, 3D scanning has allowed the architectural interrelationships of the spaces and fabric of Hawksmoor's church to be further explored, and has produced some extraordinary images. Greenwich University architectural tutor Simon Withers has for a prolonged period of time been scanning and photographing the church, accessing most of its vertiginous and secret places in his quest for completeness.

During the Second World War, on 19 March 1941, incendiary bombs hit the roof of St Alfege. Much damage was

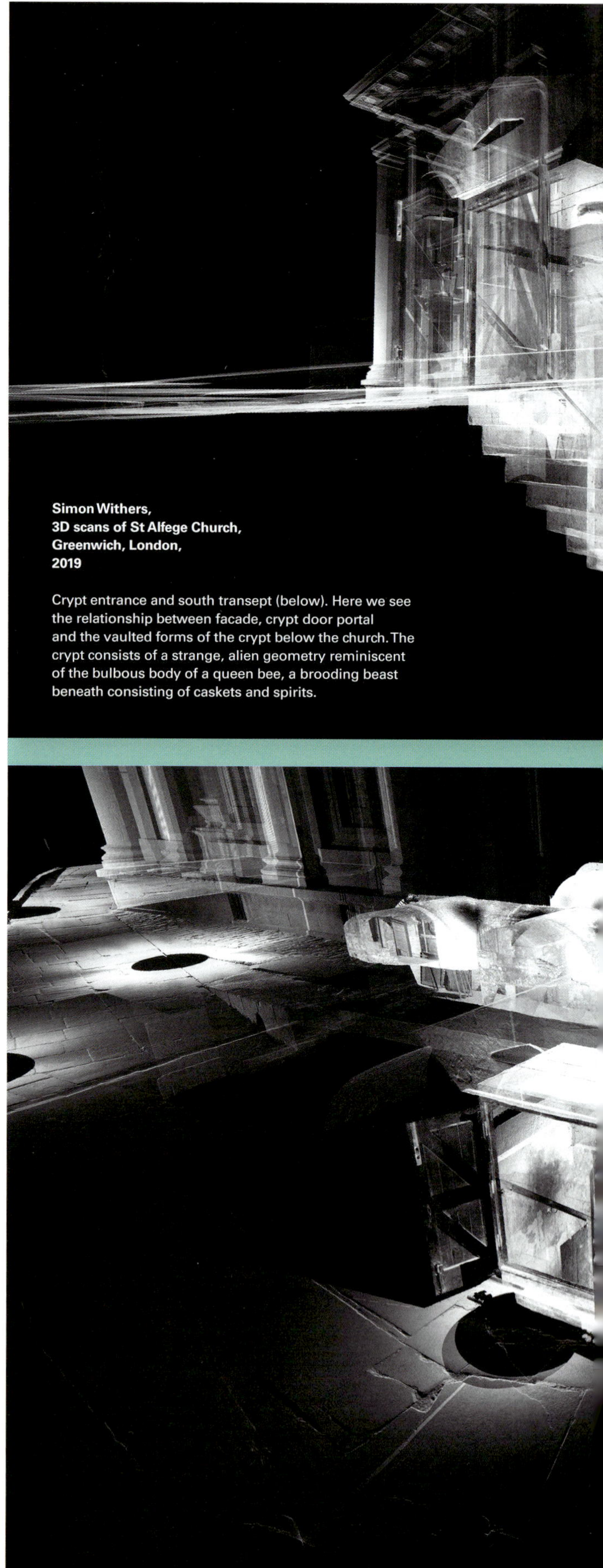

Simon Withers,
3D scans of St Alfege Church,
Greenwich, London,
2019

Crypt entrance and south transept (below). Here we see the relationship between facade, crypt door portal and the vaulted forms of the crypt below the church. The crypt consists of a strange, alien geometry reminiscent of the bulbous body of a queen bee, a brooding beast beneath consisting of caskets and spirits.

incurred; the roof collapsed, and with it Hawksmoor's oval ceiling to the nave. However, much survived, including the walls and the beautiful wooden pillars behind the altar, as well as those separating the nave from the entrance lobby. The rebuilding began in 1946, led by Professor Albert Richardson, who sought to replace as faithfully as possible what was there before, including Thornhill's trompe-l'oeil painting above the altar.

So what we see in the scans, like the fictional Hawksmoor narratives, are constellations of varying times, all being perceived simultaneously. Much like looking up at the night sky, lights from billions of different times hit our retinas all at once to create a virtual reality (some stars may have died while the light has been travelling). Indeed, the raw scans produce 'point clouds' that are akin to miniature universes.

So Hawksmoor's church still harbours secrets yet to be unearthed. The scans reveal that the great ecclesiastical ship of St Alfege is not just defined by its strict yet architecturally playful aboveground geometries; it also has a belowground underbelly that is much more expediently composed. Seeing these architectural systems juxtaposed for the first time, the geometries of the arched crypt take on a much more organic form, lurking in the darkness and shadows revealed by the scanner's all-seeing laser eye. A real beast from beneath, buried in the ancient mud of London, and a time-travelling landscape of stone and ideas. ◭

Notes
1. Peter Ackroyd, *Hawksmoor*, Hamish Hamilton (London), 1985, p 5.
2. Owen Hopkins, *From the Shadows: The Architecture and Afterlife of Nicholas Hawksmoor*, Reaktion Books (London), 2015, p 260.
3. Iain Sinclair, *Lud Heat*, Albion Village Press (London), 1975.
4. Ackroyd, *op cit*.
5. See www.st-alfege.org/Groups/317364/Thornhill_drawing.aspx.

Pierre Bélanger is a settler designer, originally from Montréal and Ottawa, based in Boston, traditional lands of the Massachusett Peoples, traditional territory of the Wampanoag and Nipmuc Nations. Together with Alexander S Arroyo, Ghazal Jafari, Pablo Escudero, Hernán Bianchi-Benguria,and Tiffany Kaewen Dang, he co-founded the anti-disciplinary non-profit organisation OPEN SYSTEMS in 2001. Their work is committed to fighting against spatial inequalities, territorial injustices and environmental inhumanities that threaten livelihoods, sovereignties and futures dependent on rivers, rains, waters, wetlands, estuaries and the oceans.

Harry Bix is the recipient of the Derek Jarman Scholarship (2014) and the Betty Mclean Award presented by Bruce Mclean (2016). He is the founder and producer of his art practice East Anglia Records, and his work has been shown at Tate Modern, the Institute of Contemporary Arts (ICA), Southwark Park Galleries, Raven Row, Charlton Gallery and SET in London, and the J:Gallery in Shanghai. He is a lecturer on landscape architecture at the University of Greenwich in London.

Neil Brenner is Professor of Urban Theory at the Harvard University Graduate School of Design (GSD). His most recent books include *New Urban Spaces: Urban Theory and the Scale Question* (Oxford University Press, 2019); *Critique of Urbanization: Selected Essays* (Birkhäuser, 2017); and the edited volume *Implosions/Explosions: Towards a Study of Planetary Urbanization* (Jovis, 2014). He is currently collaborating with Nikos Katsikis in a critical study of how geospatial data sources are used to study capitalist urbanisation, and in a longer-term project on the hinterlands of the Capitalocene.

Luis Callejas is an architect and founder of LCLA office. He was recently the Patrick Geddes Fellow at the University of Edinburgh, and taught architecture and landscape architecture at Harvard GSD. He is currently an associate professor of landscape architecture at the Oslo School of Architecture and Design.

James Corner is the founding partner of Field Operations, with offices in New York, San Francisco, London and Shenzhen. His work has been recognised with the National Design Award and the American Academy of Arts and Letters Award in Architecture, among others, and has been widely published and exhibited, including at the Museum of Modern Art (MoMA) in New York, Royal Academy of Arts in London, and the Venice Architecture Biennale. His books include *The High Line* (Phaidon, 2015), *The Landscape Imagination* (Princeton University Press, 2014), and *Taking Measures Across the American Landscape* (Yale University Press, 1996). He is a professor of landscape and urbanism at the University of Pennsylvania School of Design, and serves on the board of the Forum for Urban Design.

Teddy Cruz is Professor of Public Culture and Urbanism in the Department of Visual Arts at the University of California, San Diego, and a principal of Estudio Teddy Cruz + Fonna Forman, a research-based political and architectural practice, also in San Diego. He is known internationally for his urban research of the Tijuana/San Diego border, advancing border neighbourhoods as sites of cultural production from which to rethink urban policy, affordable housing and public space. Awards include the Rome Prize in Architecture (1991), Ford Foundation Visionaries Award (2011) and the 2018 Vilcek Prize in Architecture.

Gareth Doherty is an associate professor of landscape architecture, and Director of the Master in Landscape Architecture programmes at Harvard GSD. His research and teaching focus on the intersections between landscape architecture and anthropology. Recent research projects have centred on landscape-related practices at various sites across the postcolonial and Islamic worlds, specifically in the Arabian peninsula, West Africa, and Latin America and the Caribbean. He is the author of *Paradoxes of Green: Landscapes of a City-State* (University of California Press, 2017).

Pol Fité Matamoros is pursuing a PhD in landscape architecture and environmental planning at the University of California, Berkeley, as recipient of the Regent's Fellowship. His research focuses on the dialectical relationship between political-economic transformations and the production of space, particularly centred in the European South. His recent research collaborations include research associate for the project Atlas for a City-Region: Imagining the Post-Brexit Landscapes of the Irish Northwest and Communications Manager of the Urban Theory Lab at Harvard GSD.

Fonna Forman is Professor of Political Theory and Founding Director of the Center on Global Justice at the University of California, San Diego, and a principal of Estudio Teddy Cruz + Fonna Forman. A theorist of ethics and public culture, her work focuses on human rights, climate justice, border ethics and equitable urbanisation. She is known internationally for her revisionist research on Adam Smith, recuperating the public dimensions of his thought. She was Vice-Chair of the University of California's 2015 *Bending the Curve* report on climate change solutions, and serves on the Global Citizenship Commission, advising UN policy on human rights in the 21st century.

Matthew Gandy is Professor of Geography at the University of Cambridge. His publications include *Concrete and Clay: Reworking Nature in New York City* (MIT Press, 2002), *The Fabric of Space: Water, Modernity, and the Urban Imagination* (MIT Press, 2014) and *Moth* (Reaktion, 2016), along with articles in the *New Left Review, International Journal of Urban and Regional Research, Society and Space* and many other journals. He is currently researching the interface between the cultural and scientific aspects of urban biodiversity.

Rania Ghosn is a founding partner of DESIGN EARTH, and an associate professor of architecture and urbanism at the Massachusetts Institute of Technology (MIT). Her research examines the geographies of technological systems to prompt the debate on the environment. The work of DESIGN EARTH is in the collection of the Museum of Modern Art (MoMA) in New York, and has been exhibited at the Venice Architecture Biennale and other leading international events and museums. She is co-author, with El Hadi Jazairy, of *Geostories: Another Architecture for the Environment* (Actar D, 2018), a founding editor of the *New Geographies* journal, and editor-in-chief of *Landscapes of Energy* (Harvard GSD, 2010).

Charlotte Hansson is an architect and partner at LCLA office. Her experience in Scandinavia includes working at Space group, White Arkitekter and Alab. Most recently, LCLA's work has been exhibited at the Chicago Architecture Biennial, Venice Architecture Biennale (2018), Lisbon Architecture Triennale (2016) and Oslo Architecture Triennale (2016), among others.

El Hadi Jazairy is a founding partner of DESIGN EARTH and an associate professor of architecture at the University of Michigan in Ann Arbor. His honours include the Architectural League Prize for Young Architects, Prix Europan, and Rougerie Foundation first prize. He is co-author, with Rania Ghosn, of *Geographies of Trash* (Actar D, 2015), a founding editor of *New Geographies*, and editor-in-chief of *Scales of the Earth* (Harvard GSD, 2010).

Nikos Katsikis is an urbanist working at the intersection of urbanisation theory, design and geospatial analysis. He holds a Doctor of Design from Harvard GSD, and is currently a postdoctoral fellow at the University of Luxembourg. He is also a research affiliate of the Urban Theory Lab at Harvard, where he has also served as Instructor in Urban Planning and Design.

Christina Leigh Geros is an architect, landscape architect and urban designer specialising in design-led research that critically engages the production of knowledge infrastructures related to climate- and neuro-ecologies. She is currently a research fellow with Monsoon Assemblages and a tutor in the MA Environmental Architecture Programme at the Royal College of Art (RCA), leading The Orang-orang and the Hutan research project. Previously based in Jakarta, her work as the Design Director of Anexact Office and Design Research Strategist for PetaBencana.id informs her current practice of designing engagements, implementations and interfaces of investigation that bridge across platform, scope and inquiry.

Kate Orff is the founder of SCAPE, a 50-person landscape architecture and urban design practice based in New York City and New Orleans, Louisiana. She is also the director of the MS in Architecture and Urban Design (MSAUD) programme at Columbia University's Graduate School of Architecture, Planning and Preservation (GSAPP) and co-director of the Center for Resilient Cities and Landscapes. SCAPE's Public Sediment for Alameda Creek project was created by a team led by design principal Gena Wirth in collaboration with Arcadis, the Dredge Research Collaborative, TS Studio, UC Davis Department of Human Ecology and Design, Cy Keener and Architectural Ecologies Lab.

Toya Peal is a landscape architect with a particular interest in the relationship between people and place and the arts. She is interested in creating places through the design and curation of space through music, events, arts and architecture. Since completing her Masters at the University of Greenwich, she has worked in London and Hong Kong, and now works for a landscape architecture practice in Edinburgh.

Neil Spiller is Editor of △D, and was previously Hawksmoor Chair of Architecture and Landscape and Deputy Pro Vice Chancellor at the University of Greenwich, London. Prior to this he was Vice Dean at the Bartlett School of Architecture, University College London (UCL). He has made an international reputation as an architect, designer, artist, teacher, writer and polemicist. He is the founding director of the Advanced Virtual and Technological Architecture Research (AVATAR) group, which continues to push the boundaries of architectural design and discourse in the face of the impact of 21st-century technologies. Its current preoccupations include augmented and mixed realities and other metamorphic technologies.

Tiago Torres-Campos is a Portuguese landscape architect and an associate professor at the Rhode Island School of Design in Providence. He has been published internationally and founded CNTXT Studio, a research-by-design platform focusing on the study of landscape and its intersections with architecture, art, design and digital media. He is currently completing a PhD in Architecture by Design, investigating landscape and architectural modes of thinking geologically and their applications to contemporary design practices.

Tim Waterman is a senior lecturer in landscape architecture history and theory at the Bartlett School of Architecture, UCL. He is at work on the book *Landscape Citizenships*, and has recently co-edited two others: *Landscape and Agency: Critical Essays* (Routledge, 2017) with Ed Wall; and the *Routledge Handbook of Landscape and Food* (Routledge, 2018) with Joshua Zeunert.

What is *Architectural Design*?

Founded in 1930, *Architectural Design* (△) is an influential and prestigious publication. It combines the currency and topicality of a newsstand journal with the rigour and production qualities of a book. With an almost unrivalled reputation worldwide, it is consistently at the forefront of cultural thought and design.

Each title of △ is edited by an invited Guest-Editor, who is an international expert in the field. Renowned for being at the leading edge of design and new technologies, △ also covers themes as diverse as architectural history, the environment, interior design, landscape architecture and urban design.

Provocative and pioneering, △ inspires theoretical, creative and technological advances. It questions the outcome of technical innovations as well as the far-reaching social, cultural and environmental challenges that present themselves today.

For further information on △, subscriptions and purchasing single issues see:

http://onlinelibrary.wiley.com/journal/10.1002/%28ISSN%291554-2769

Volume 89 No 1
ISBN 978 1119 453017

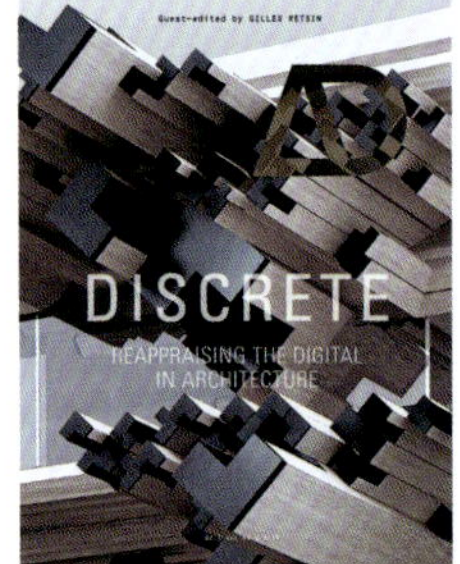

Volume 89 No 2
ISBN 978 1119 500346

Volume 89 No 3
ISBN 978 1119 546023

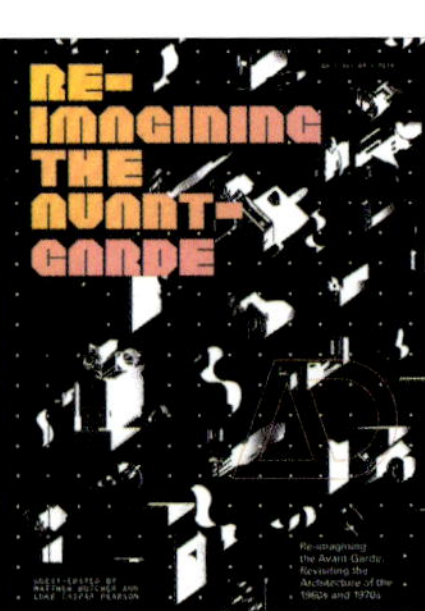

Volume 89 No 4
ISBN 978 1119 506850

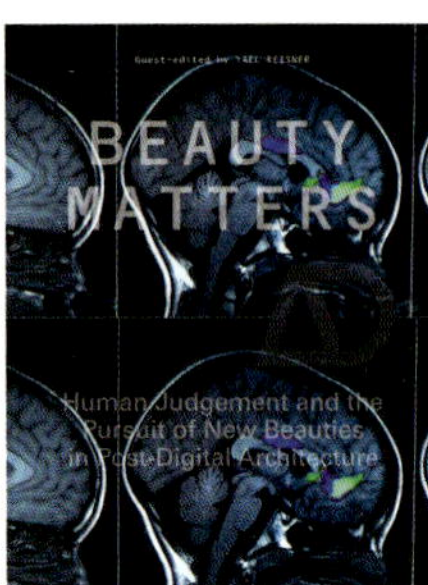

Volume 89 No 5
ISBN 978 1119 546245

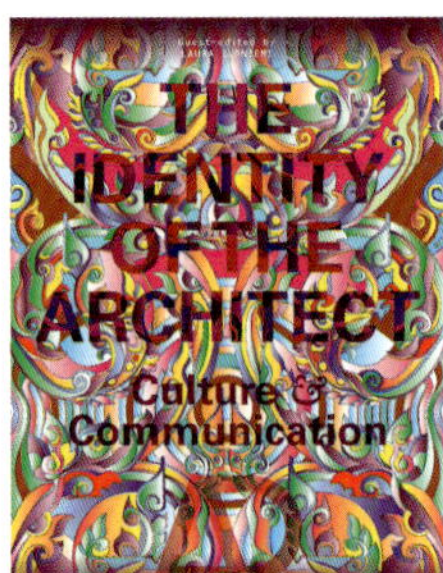

Volume 89 No 6
ISBN 978 1119 546214